Physics Workbook Vo.

Revised Spring 2021

The solutions are available <u>for free</u> as a pdf at http://www.robjorstad.com/Phys161/161Workbook.htm.

Use a laptop, tablet, or phone to access the solutions at my website above.

Find the chapter you want then open that link on your device.

The questions (this book) and the answers (on your device) ready to go at the same time!

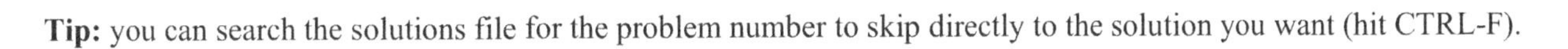

Tip: you can search the solutions file for the problem number to skip directly to the solution you want (hit CTRL-F).

Do not attempt any demonstrations described in this book. Some are extremely dangerous and can only be successfully performed by trained & experienced professionals (i.e. Bed of nails sandwich with sledgehammer).

WORK & ENERGY

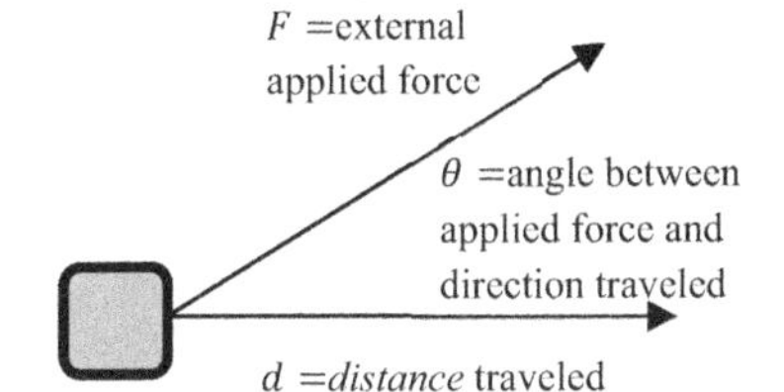

For a constant force work is $W = Fd\cos\theta = \pm F_{\parallel}d$.
Use the – for a force component anti-parallel to *displacement*.
If you are into dot products: $W = \vec{F}\cdot\Delta\vec{s}$ where $\Delta\vec{s}$ is the 3D displacement vector.

For a non-constant force use calculus:

$$W = \vec{F}\cdot\Delta\vec{s} = \begin{cases} \int F_x dx + \int F_y dy + \int F_z dz \\ \\ \int F\cos\theta\, ds \end{cases}$$

Work done by a force on an object is + if that force tends to speed that object up.
The net work by all external forces on an object is equal to the CHANGE in kinetic energy of that object.

$$\Delta KE = W_{all\ external\ forces}$$

Some forces in nature are conservative. A great example of a conservative force is a spring. When you compress a spring, the spring pushes on your hand and does negative work (force is opposite the direction of motion). As you release the spring, it pushes in the same direction as motion and does positive work. After returning to the initial position the spring has done no NET work. The initial energy is unchanged; energy is said to be conserved.

Deriving the Work-Energy Theorem
The Work-Energy Theorem is $W_{net} = \Delta K$. Perhaps you are curious where this comes from?

Start from the mathematical definition of work and see what happens. The definition of work is given as

$$W = \int_i^f \vec{F}\cdot d\vec{s}$$

If we restrict ourselves to motion only in the *x*-direction $d\vec{s} = dx\hat{\imath}$

$$W = \int_i^f (F_x\hat{\imath})\cdot(dx\hat{\imath}) = \int_i^f F_x dx$$

Now use $F_x = ma_x = m\frac{dv_x}{dt}$. When you plug it in you get

$$W = \int_i^f m\frac{dv_x}{dt}dx$$

Next comes a subtle step. We are free to rearrange the order of differentials inside the integrand. This gives

$$W = \int_i^f m\frac{dx}{dt}dv_x = \int_i^f mv_x\, dv_x$$

From there the derivation is straightforward. If you want you can do the same thing in 3D. At some point you'll need to use $v^2 = v_x^2 + v_y^2 + v_z^2$.

WATCH OUT! It may seem we switch between distance & displacement quite often.
For an object moving in one direction (with no reversals)

$$displacement = distance \times direction$$
$$\Delta\vec{s} = d(some\ unit\ vector)$$

Practical Work & Energy Comments

For any constant external force

$$Work = Fd\cos\theta$$

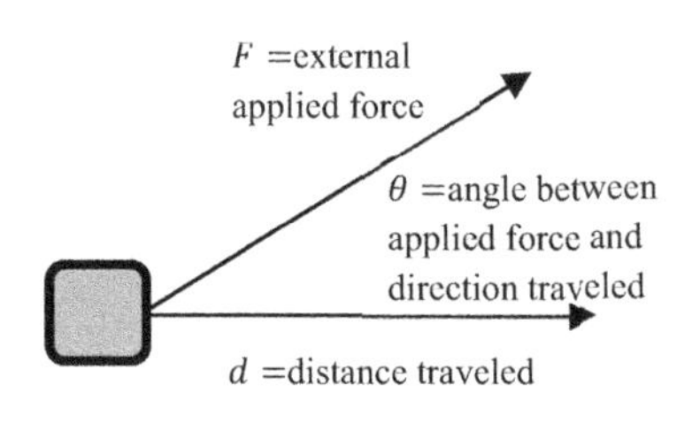

Example:

A block is pulled with constant tension $T = 12.0$ N at an angle of 10.0° above the horizontal. The block never loses contact with the surface. Determine the work done by the tension as it pulls the block 3.00 m to the right.

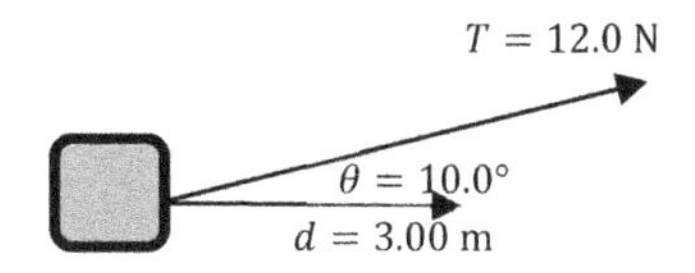

$$W_{Tension} = (12.0\text{N}) \cdot (3.00\text{m}) \cdot \cos 10.0° = 35.5\ \text{N} \cdot \text{m} = 35.5\ \text{J}$$

Here the new unit is J = Joules…a real gem of a unit.

Another option for any constant external force is to use the dot product. Here $\Delta\vec{r}$ is the displacement vector. Practically speaking, one splits the force into *x*, *y*, and *z* components. After splitting up the force into components one can use

$$W = \vec{F} \cdot \Delta\vec{r} = F_x\Delta x + F_y\Delta y + F_z\Delta z$$

Redoing Example:

A block is pulled with constant tension $T = 12.0$ N at an angle of 10.0° above the horizontal. The block never loses contact with the surface. Determine the work done by the tension as it pulls the block 3.00 m to the right.

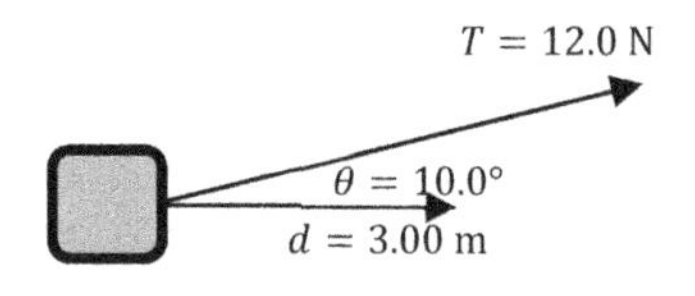

First split up the force into x and y components:

$$\vec{T} = (12.0\cos 10.0°\,\hat{\imath} + 12.0\sin 10.0°\,\hat{\jmath})\text{N} = \left(11.\underline{8}2\hat{\imath} + 2.0\underline{8}4\hat{\jmath}\right)\text{N}$$

Next, I determine the *displacement vector* is

$$\Delta\vec{r} = (3.00\hat{\imath} + 0.0\hat{\jmath})\text{m}$$

$$W_{Tension} = T_x\Delta x + T_y\Delta y + T_z\Delta z$$

$$W_{Tension} = \left(11.\underline{8}2\text{N}\right)(3.00\text{m}) + \left(2.0\underline{8}4\text{N}\right)(0.0\text{m}) + (0.0\text{N})(0.0\text{m}) = 35.5\ \text{J}$$

7.1 A block of mass m is pushed up an incline of angle θ by a horizontal force $\vec{F}$. A kinetic coefficient of friction μ exists between the block and the plane. The block slides a distance L up the incline. The magnitude of the acceleration due to gravity is g.

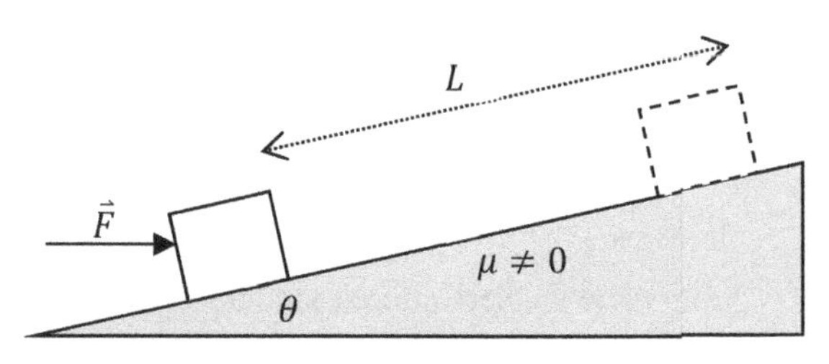

a) Draw the FBD of the block on the plane.
b) Determine an expression for the normal force in terms of m, g, F, and θ. Note: $n \neq mg\cos\theta$ in this case!
c) Determine an expression for the friction force in terms of θ, m, g, F, and μ.
d) Determine an algebraic expression for the work done by the force F in terms of F, L, and θ.
e) Determine the work done by gravity in terms of m, g, L, and θ.
f) Determine the work done by friction in terms of L, θ, m, g, F, and μ.
g) Determine the work done by the normal force.
h) Assume the block starts from rest. Use $W_{net} = \Delta K$ to determine the final speed of the block.

In general, forces that *would tend* to speed you up do positive work, forces that *would tend* to slow you down do negative work. Friction *usually* does negative work but not always. Consider the next problem...

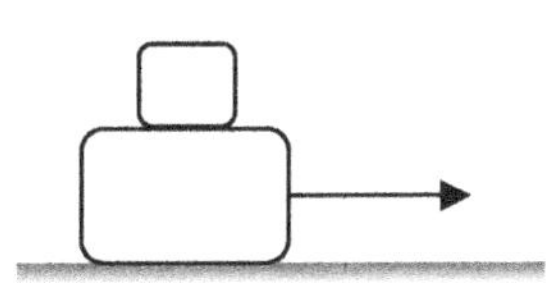

7.2 Two blocks are stacked on top of each other. There is no friction between the bottom block and the floor. There is friction between the top block and the bottom block. The system is pulled to the right by force applied to the lower block. The top block does not slide relative to the lower block.

a) Does friction do positive, negative, or no work on the top block?
b) Does friction do positive, negative, or no work on the bottom block?
c) Does friction do positive, negative, or no work on the two block system?
d) How do the above answers change if the top block was pulled to the right instead of the bottom block? Again, assume the top block does not slide relative to the lower block.

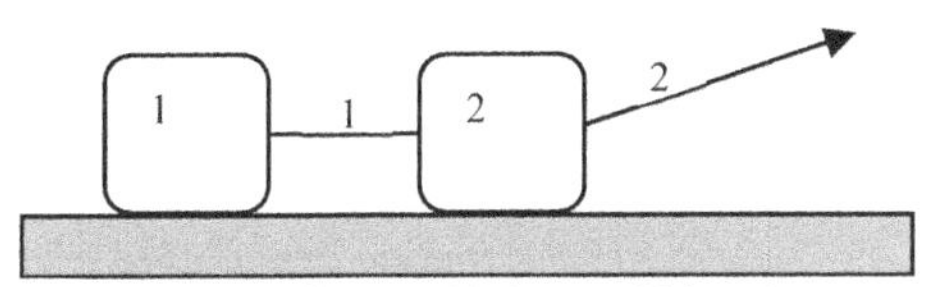

7.3 Two identical blocks, each mass m, blocks are connected by a light, inextensible string on a horizontal surface. A zombie pulls the blocks to the right using an string (also massless and inextensible) with angle θ. The blocks start from rest. After the zombie has walked distance d the block chain turns into a bit coin...er...I mean the blocks are each moving with speed v. Assume the coefficient of kinetic friction between each block and the floor is μ. For ease of communication I numbered the blocks and strings.

a) Is the work done by string 1 on the block 2 zombie positive, negative, or zero?
b) Is the work done by string 1 on the block 1 from the zombie positive, negative, or zero?
c) Is the net work done by string 1 on the *two block system* positive, negative, or zero?
d) Do your above results change if the two masses are *not* identical?
e) Is the frictional force exerted by the ground on the trailing block the same magnitude as the frictional force from the ground on the leading block? Explain why or why not?
f) Determine the magnitude of tension in string 2 in terms of θ, d, v, g, m, and μ.
g) **Challenge:** Assume instead the tension is string 2 has magnitude T_2 but unknown angle. Determine the angle in terms of T_2, d, g, m, and μ.
h) **Challenge:** now assume a more general case where the block farther from the zombie is mass 2 and the one closer to the zombie is mass 1. Determine the angle which gives the block the greatest speed after travelling distance d. Assume string 1 has a constant tension magnitude regardless of angle. After determining this angle, determine the string 1 tension magnitude required to give the blocks speed v after traveling distance d with the string 1 tension applied at the optimum angle.

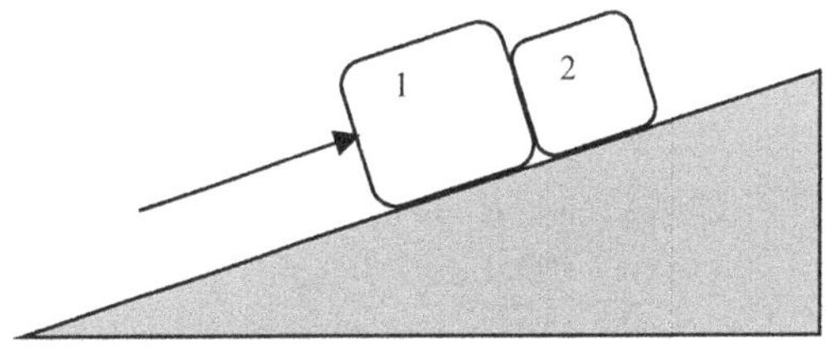

7.4 A zombie lowers two blocks down an incline of angle θ. The block closer to the zombie has mass m while the smaller block has 25% less mass. The zombie pushing the blocks pushes parallel to the incline with magnitude P. The process starts with the blocks at rest. The zombie's mind, whatever is left of it, contains a nano-bot which releases a psychotropic chemical causing her to stop pushing on the blocks once they reach speed v. Assume the coefficient of kinetic friction between each block and the floor is μ. Assume μ is small but not entirely negligible.

a) Is the work done by normal force (between the blocks) on the block 1 positive, negative, or zero?
b) Is the work done by normal force (between the blocks) on the block 2 positive, negative, or zero?
c) Is net work done by normal force (between the blocks) on the *two block system* 1 positive, negative, or 0?
d) Determine the sign of each non-zero work term.
e) How far will the zombie push the blocks before nano-bot releases the chemical?

7.5 Car of mass m drives on a banked curve in a circular path of radius R with speed v. The curve is angled at θ above the horizontal. The car is driving at the maximum speed possible without sliding to the outside of the track. The coefficients of friction between the tires and the road are $\mu_s = 0.9$ & $\mu_k = 0.7$. Be sure to determine all three answers before looking at the solution.

a) Determine the work by the normal force as the car drives halfway around the track.
b) Determine the work done by gravity as the car drives halfway around the track.
c) Determine the work done by friction as the car drives halfway around the track.

Work for non-constant force acting along the x-direction is given by

$$W = \int_i^f \vec{F} \cdot d\vec{s} = \int_i^f (F_x\hat{\imath}) \cdot (dx\hat{\imath}) = \int_i^f F_x dx$$

7.6 The force caused by a spring increases as you stretch the spring more. This is a non-constant force directed along the x-axis. The equation for the force caused by the spring is given by

$$\vec{F}_{spring} = -k\Delta\vec{x}$$

where k is called a spring constant and $\Delta\vec{x}$ is the displacement of the spring from the unstretched position.

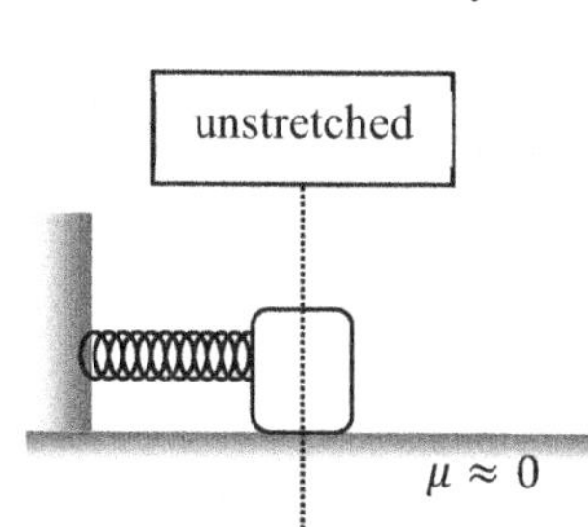

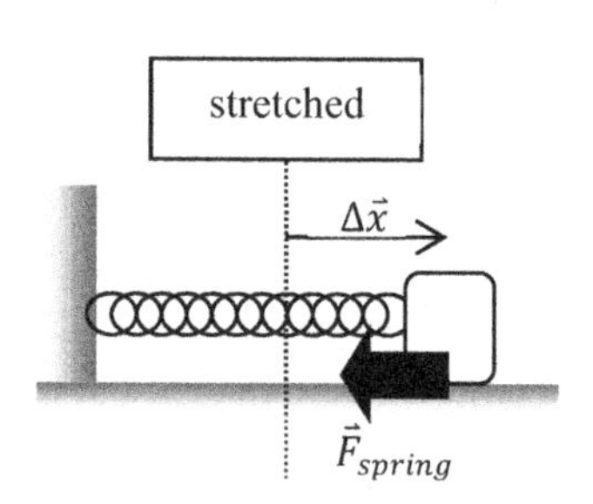

If we set the unstretched position to $x = 0$ we may write the force as

$$\vec{F}_{spring} = -k\vec{x} \qquad \text{(vector: position } \vec{x} \text{ can be } \pm)$$

In free body diagrams *the magnitude* of the spring force is

$$F_{spring} = kx \qquad \text{(magnitude: distance } x \text{ is always } +)$$

The arrow associated with kx will indicate whether or not to include a minus sign in the force equation.

a) Assume a spring has k = 40.0 N/m. Verify the units are correct.
b) Determine the work done by the spring as the object is pulled from the unstretched position to $\vec{x}$ = 6.0 cm$\hat{\imath}$.
c) Determine the work done by the spring as the object is pulled from $\vec{x}$ = 6.0 cm$\hat{\imath}$ to $\vec{x}$ = 12.0 cm$\hat{\imath}$.
d) The mass is released from rest when it is located at $\vec{x}$ = 12.0 cm$\hat{\imath}$. The mass moves to the left until coming momentarily to rest at $\vec{x}$ = −12.0 cm$\hat{\imath}$. Determine the work done by the spring. Explain…

Watch out: for a horizontal springs equilibrium and unstretched positions are the same. For a vertical they are not.

7.7 Consider a bullet in the barrel of a gun. The force on the bullet changes as the bullet travels down the length L of the barrel. The mass of the bullet is m. Suppose the net force on the ball is modeled by

$$F_x = \alpha x - \beta x^2$$

where α and β are constants. The bullet starts from rest slightly to the right of the origin ($x_i \approx 0$).

a) Determine the units for α and β.
b) Determine the work done by this force as the bullet travels the length of the barrel in terms of α, β, and L.
c) Use $W_{net} = \Delta K$ to determine the speed of the bullet as it leaves the barrel in terms of α, β, m, and L.
d) At what non-zero distance is there no net force on the bullet? Answer in terms of α and β.
e) Barrels longer than the answer to part d would be counterproductive if our model is correct. Explain why.
f) Do you think this is a decent model for the force on the bullet? Defend your answer. If not, how might you modify the model to make it more realistic?

7.8 An electron is trapped in a 1D quantum electronic structure. The electron can move…but only in one dimension. By adjusting the material composition of the well, an engineer has created a situation where the net force acting on the electron is given by

$$F(x) = \frac{\alpha x}{\beta^2 + x^2}$$

where α and β are constants. Notice this force looks a lot like a spring force when $x \ll \beta$.

a) Determine the units on α and β.
b) Determine the work done on the electron as it moves from $x_i = 0$ to $x_f = L$.

Note: while this *particular* force is purely fictional, quantum structures are used in electrical engineering and condensed matter (solid state) physics. Quantum wells restrict the motion of electrons to 2D motion and quantum wires to 1D. Quantum dots act a lot like really big atoms and effectively lock the electron in place…in some sense you could think of this as a 0D situation. Pretty cool stuff, tons of practical applications. If you dig it, consider going into electrical engineering or condensed matter physics.

Note^2: Something that interested me back in grad school was solar technology and quantum dots. I read once that scientists had proposed using quantum dots inside a solar cell to improve the efficiency of energy collection. The quantum dots, which act like really big atoms, absorb photons just like atoms. Due to their unique structure, quantum dots can absorb low energy photons that are normally not collected by solar cells. You might find it interesting to do a web search for "quantum dot solar cell" and see how the technology is evolving.

Power

Instantaneous power is given by

$$\mathcal{P} = \frac{dE}{dt}$$

To be clear this is the rate of energy change over time. Notice that we can get an *average* value using

$$\mathcal{P}_{avg} = \frac{\Delta E}{\Delta t} = \frac{Work}{time}$$

Note: it can be shown that the instantaneous power is also given by

$$\mathcal{P}_{inst} = \vec{F} \cdot \vec{v}$$

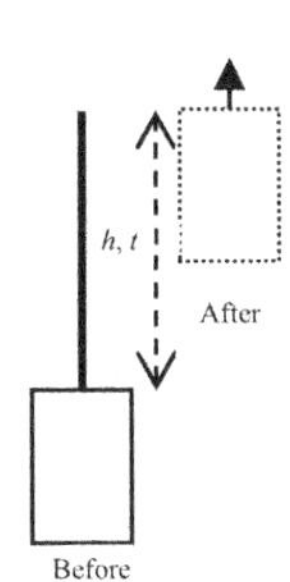

7.9 An elevator of mass m accelerates upwards from rest. It travels distance h in time t.

a) Determine the acceleration in terms of h and t.
b) Determine the final speed in terms of h and t.
c) Determine the tension in the cable in terms of m, h, t, and g.
d) Determine the *average* power delivered by the motor.
e) Determine the *instantaneous* power delivered by the motor as the object reaches its final position.

7.10 Requires Derivatives: The position of a particle as a function of time is given by

$$x(t) = 2.00 - 3.00t^3$$

The mass of the particle is $m = 0.250$ kg.

a) Determine the units appropriate for 2.00 and 3.00.
b) Determine $v(t)$ and $a(t)$.
c) Write an expression for the net force acting on the particle.
d) Determine the *instantaneous* power associated with the net force at time $t = 2.00$ sec.
e) Determine the *average* power between $t = 0$ and 2.00 sec.

Why do I really like this problem? Imagine you took a video of something moving. For example, suppose we took a video of a hummingbird in flight. You could use video capture techniques to track the hummingbird and get an estimate of its position versus time. From that data, a trendline could give you an approximate equation for position as a function of time. From that we could learn about the average & instantaneous power output of the hummingbird!

7.11 A car of mass m drives on flat ground with constant speed v. Air resistance is non-negligible. The power output of the engine is $\mathcal{P}_0$. The car then drives up a hill inclined θ above the horizontal at the same speed v. Assume the force of air-resistance is unchanged since the car is still travelling at the same speed. By what factor must the power output of the car increase? Answer in terms of m, g, v, θ and $\mathcal{P}_0$.

7.12 A car of mass $10\underline{0}0$ kg accelerates from rest in two stages. During the first stage the car goes from 0 to $20.0\frac{\text{m}}{\text{s}}$ in 4.00 seconds. During the second stage the car goes from $20.0\frac{\text{m}}{\text{s}}$ to $40.0\frac{\text{m}}{\text{s}}$ in 6.00 seconds. Assume 100% of the car engine's power output is successfully used to accelerate the car. Determine the *average* power output of the engine for the entire 10.00 seconds of motion. Note: if you are interested in learning about engine efficiencies, this is discussed when we do thermodynamics...typically in your second or third semester physics course.

7.13 Good review A powerlifter of mass m stands on a scale holding mass m in front of her face. The woman raises the mass above her head using a three stage process. In the first stage, the woman accelerates the mass to speed v. In the second stage, she raises the mass at constant speed v. In the third stage, she slows the mass to a stop at the same constant rate as in the first stage. The stages each cover equal distances. For ease of communication, let us say the force the woman exerts on the extra mass is $\vec{F}$. Notice this $\vec{F}$ will change over time.

a) Sketch what the following plots should look like. Concern yourself with getting the shapes of the curves correct. By this I mean when are the plots sloped, curving, flat, positive/negative. Note: for all plots involving vectors assume up is the positive direction.
   i. $\vec{x}\ vs\ t$
   ii. $\vec{v}\ vs\ t$
   iii. $\vec{a}\ vs\ t$
   iv. $\vec{F}\ vs\ t$
   v. Scale reading versus time $\vec{n}\ vs\ t$
   vi. $K\ vs\ t$...here K is kinetic energy
   vii. $\mathcal{P}\ vs\ t$...here $\mathcal{P}$ is the power delivered to the mass by the woman's applied force

b) Now assume the total distance traveled is $h = 45.0$ cm, the max speed attained is $v = 1.00\frac{\text{m}}{\text{s}}$, and the mass is 60.0 kg. Sketch plots with appropriate numerical values and labels. Hint: first determine the magnitude of the acceleration and the time for each stage.

c) Now think: how would the plots be different if the woman lowered the masses using a similar three-stage process. In particular, which graphs remain unchanged, which merely flip signs, and which do something else?

7.14 Requires separation of variables: Constant power does <u>not</u> imply constant acceleration. Consider a car of mass m travelling under constant power $\mathcal{P}$. The car starts from rest. After travelling time t the speed is v.

a) Determine v in terms of the other givens.

b) Use your expression to determine $x(t)$ and $a(t)$.

c) Determine the ratio of the car's speed after distance $2d$ to the car's speed after distance d.

d) Consider a constant *acceleration* truck that starts from rest and reaches the same speed v after time t. Determine the ratio of the truck's speed after distance $2d$ to the truck's speed after distance d.

e) A common physics demonstration device is a fan cart that rolls along a metal track. The cart can be modeled as moving with negligible friction. Devise a method to determine if the car is best modeled as having constant acceleration or constant power.

f) A high performance car of mass 400 kg ($\approx$ 880 lbs) delivers 220 kW of power ($\approx$ 340 hp) to the wheels. Plot $x(t)$, $v(t)$, and $a(t)$ assuming a constant power model is valid. Discuss the validity of the model after looking at the values on the plot.

Force-Distance Trade-offs

For constant forces we know

$$W = Fd\cos\theta = F_{\parallel}d$$

Notice a small force can do the same amount of work as a large force as long as the force is applied over a longer distance. This is called a force-distance trade-off.

7.15 Consider raising box of mass m initially at rest on the ground. The goal is to ultimately get the box on the loading dock distance h above the ground. Note: I will assume the box is always lifted at constant speed to simplify the problem. Strictly speaking this cannot be true since the box starts at rest, then moves upwards, and finally comes to rest on the loading dock. If the box is moved slowly, however, the acceleration phases (speeding up and slowing down) are probably negligible compared to the rest of the problem.

a) One person lifts the box straight up at constant speed. Using this style, what applied force must be used to lift the block at constant speed?

b) Another person pushes the box horizontally up a 19.5° ramp of length L. For now, let us assume the ramp is frictionless. What *horizontally* applied force is required to raise the box at constant speed with the ramp?

c) How much work is done in by the applied force in each case? Discuss the force-distance trade-off.

A machine is a device used to change the direction and/or magnitude of a force. Do a web search for "simple machine" and you will often find six types described.

1. Screw
2. Lever
3. Wheel & axle
4. Wedge
5. Inclined plane
6. Pulley

More complicated machines can be made from these simple ones. As an example, a wedge is essentially two inclined planes put together back-to-back. As an exercise, imagine a way you could reduce the required force to raise a small mass using each of the above machines.

This is a good time to revisit the pulley problems in chapter 5. An understanding of torque in Chapter 10 will help in understanding the force-distance-trade-offs involved in levers. Some examples are on the next page.

7.16 You lift up a bowling ball to height h then put it back down. The ball has mass m. Ignoring the brief acceleration stages, the entire process happens as roughly constant speed.

a) How much work did you do on the bowling ball?
b) How much work was done by gravity while the ball was lifted and put back down?
c) Explain why energy used by your body (calories expended) and work done on the ball are not the same.

7.17 A cart of mass m is constrained to translate along a track. You tie a rope of length L between the cart and a pipe at the end of the track. Midway between the cart and the pipe you push sideways on the rope with constant force $\vec{F}$. The cart may be modeled as sliding to the right with no friction.

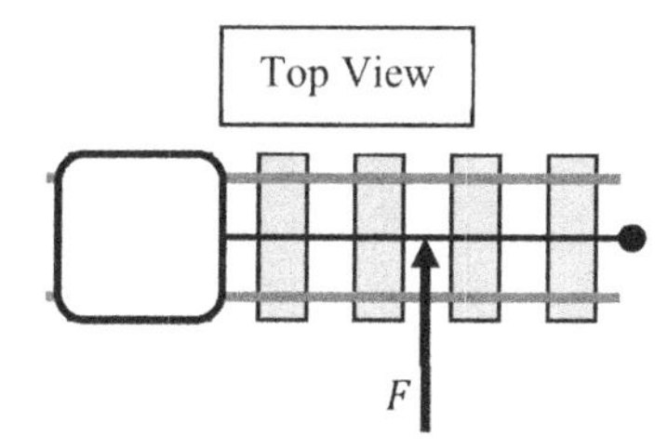

a) How far does the cart travel if the rope is pushed sideways distance $\frac{L}{5}$?
b) Determine the speed of the cart after pushing distance $\frac{L}{5}$.
c) How does this speed compare to not using any rope and simply pushing the cart to the right with force $\vec{F}$ the same distance $\frac{L}{5}$?
d) Is there any practical application of this?

7.18 A cart of mass m is constrained to move along a track with a pulley system as shown. The cart is initially distance s from the wall. Starting from rest, a winch exerts a constant tension force (magnitude T) on the cable. This causes the cart to slide from point **A** to point **B** distance d apart. You observe the cart never lifts off the track. The cable from the winch is distance h above the point where the string connects to the cart. Assume friction is negligible.

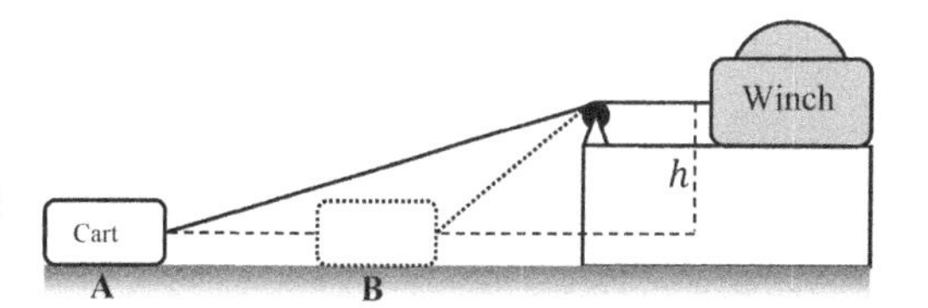

a) Determine the speed of the cart when it reaches point **B**.
b) **Challenge:** How does the answer change if kinetic friction is present with coefficient μ_k? What additional information is required to determine the result?

Using Potential Energy Instead of Work

Gravity and spring forces conserve energy. This means gravity and spring forces do no **net** work for any closed path. A closed path is one that starts and stops at the same spot. An example of a closed path would be if you moved to the right 3.00m and then come back to your starting point. Or go up and then back down the same amount. Or walk around a track in a circle and come back to the starting line.

For conservative forces $Work = -\Delta U$. Here I want to stress ΔU is CHANGE in potential energy. Gravity is one such force. For the figure at right the following logic thus holds:

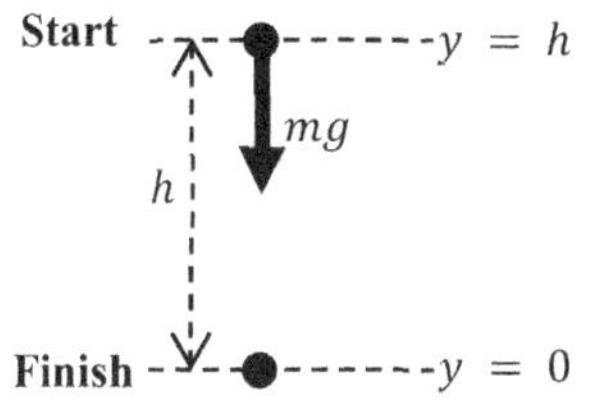

$$\Delta U_{Grav} = -(W_{Grav})$$
$$U_f - U_i = -(mgh)$$
$$U_f - U_i = -mgh$$

Start $y = h$, mg, h, Finish $y = 0$

At this point we are given a choice. We can arbitrarily choose where to set gravitational potential energy as zero. This is called our reference level. Almost always we will choose our reference level as $y = 0$ (the lowest point in the problem) to minimize the minus signs in our equations. Using this convention, our above work reduces to

$$U_{y=0} - U_{y=h} = -mgh$$
$$0 - U_{y=h} = -mgh$$
$$U_{y=h} = mgh$$

This is only valid near the earth's surface as we will learn later. In astronomical gravitation problems and some static electricity problems we often set $y = \infty$ as our reference level with zero potential energy. The gravitational potential energy becomes $U = -\frac{GmM}{r}$.

Note: the gravitational energy is shared between the earth and the ball. Because the earth has so much mass, it has essentially no acceleration or speed after the ball falls. For this reason we often misspeak and say the ball has gravitational energy. In this chapter saying things that way works out just fine. In Chapter 13 we handle astronomical objects when the sharing of the gravitational potential energy becomes important.

Conservation of Energy ($E_i = E_f$)

I usually start out energy problems by writing

$$E_i + W_{non-con} = E_f$$

If $W_{non-con} = 0$ we get

$$E_i = E_f$$

Any time the initial quantity of *something* equals the final quantity of *something* we say "*something* is conserved". Conservation laws are useful because they gives us equations. Once you have an equation, you have a model for predicting behavior. Once you have predictions, you can test your predictions with experiment which is what science is all about! Some examples of quantities with conservation laws are energy, momentum, angular momentum, charge, mass, strangeness, charm, etc. The last two are properties of quarks.

Energy/Money Analogy
Suppose you have both a checking and a savings account at a bank.
You have some money in both of them and consider both accounts as "your system".
View your savings account as potential energy and your checking as kinetic energy.
In this system, transfers between your savings & checking accounts are internal.
You still have the money somewhere in your system after the transfer.
Each time you make a purchase, money(energy) is leaving your system.
Each time you make a deposit, energy(money) is added to your system.

Method 1 (Chapter 8 Style)
The choice of system for method 1 is everything enclosed in the thick dotted line.

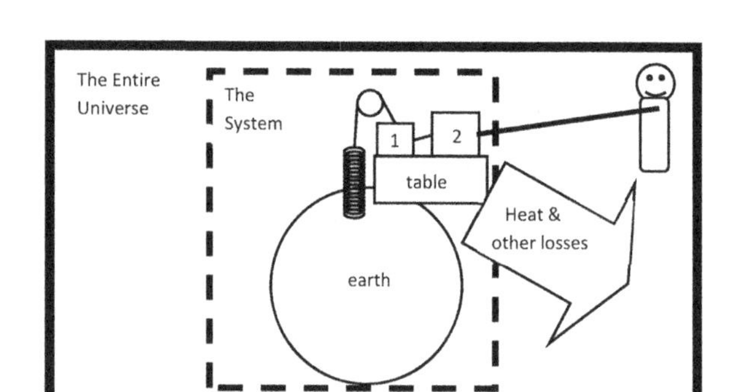

$$E_i + W_{non-con} = E_f$$

or

$$W_{non-con} = \Delta E$$

Notice the earth and the spring are inside the system (they are internal forces) and do no net work.
The force of gravity and the spring will be accounted for using SPE and GPE *instead of using work.*
The strings completely inside the system are internal and do no net work on the system.
Here $W_{non-con} = W_{friciton} + W_{space\ zombie\ tension}$
Typically you would draw an FBD and use $W = Fd\cos\theta$ to determine each term in $W_{non-con}$.

Method 2 (Chapter 7 Style)
Here the system only includes the two blocks and the string between them. Ignoring tension in the string between the blocks (it is internal) we use

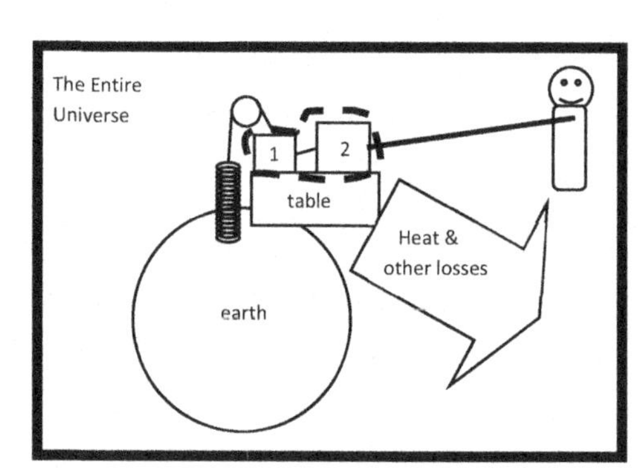

$$W_{net} = \Delta K$$

$$W_G + W_S + W_f + W_{s.z.t.} = \Delta K = \frac{1}{2}mv_f^2 - \frac{1}{2}mv_i^2$$

where $W_{s.z.t.} = W_{space\ zombie\ tension}$.
You would draw an FBD and use $W = Fd\cos\theta$ to determine each term in $W_{non-con}$.
The force of gravity and the spring are accounted for using work *instead of using SPE and GPE.*

Summary on next page...

What are the take-aways from all this blather:

- In the entire universe $E_i = E_f$. Unfortunately, if the entire universe is our system it is usually impractical to calculate anything useful.
- In many real life systems $E_i + W_{non-con} = E_f$. In these systems energy is either entering or leaving the system through the non-conservative work.
- Doing energy problems use ***either*** $W_G + W_S$ ***or*** GPE and SPE ***but not both***.

For conservative forces physicists define CHANGE in potential energy using

$$\Delta U = -Work$$

Note: friction is not a conservative force.

Typical energies are thus written as

Translational Kinetic Energy	Rotational Kinetic Energy	Spring Potential Energy	Gravitational Potential Energy
$TKE = \frac{1}{2}mv^2$	$RKE = \frac{1}{2}I\omega^2$	$U_S = SPE = \frac{1}{2}kx^2$	$U_G = GPE = mgh$

FOR PROBLEMS WITH SLIDING FRICTION, GRAVITY AND SPRINGS

Chapter 7 style: Use the Work-Energy Theorem ($W_{net} = \Delta K$). Get work *from every force* acting on the object..

$$W_G + W_S + W_f = \Delta K = \frac{1}{2}mv_f^2 - \frac{1}{2}mv_i^2$$

Chapter 8 style: Use potential energy *instead of work* for *conservative* forces (often springs & gravity…see **8.28**)

$$W_{non-conservative} = \Delta E$$

$$W_{friction} = \Delta K + \Delta U$$

$$W_{friction} = (K_f - K_i) + (U_{Gf} - U_{Gi}) + (U_{Sf} - U_{Si})$$

$$K_i + U_{Gi} + U_{Si} + W_{friction} = K_f + U_{Gf} + U_{Sf}$$

$$\frac{1}{2}mv_i^2 + mgh_i + \frac{1}{2}kx_i^2 + W_{friction} = \frac{1}{2}mv_f^2 + mgh_f + \frac{1}{2}kx_f^2$$

Watch out! In special cases friction will do no work or even positive work. If you have an external tension, add in a $W_{tension}$ term.

Technical note: gravitational energy is shared between the earth and the object and is stored in the gravitational field. Strictly speaking, a ball by itself has no gravitational energy. If you are including gravitational energy in your system, you have technically included the earth in your system.

8.1 & 8.2 Consider a simulation. One is discussed in the supplemental handouts.

8.3 Consider a block of mass m sliding on a frictionless ramp shaped like a parabola. The block has negligible size so we may consider the center of mass position of the block as being, essentially, on the ramp. The block is released from rest at initial height h.

a) Determine the max speed of the block when it reaches the bottom of the ramp.
b) At what height above the ground does the block have half of its max speed?
c) Compare the speed of the block when it is $h/2$ above the bottom versus at the bottom of the ramp. Use a ratio to clearly communicate how the two compare.

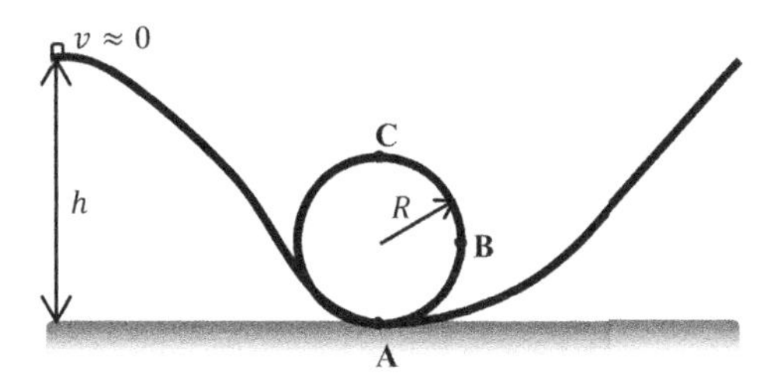

8.4 A block of negligible height slides without friction on the track shown in the figure. The block is given a very slight push to get it moving. Assume the initial velocity is close enough to zero to be ignored.

a) Determine the speed of that particle at points **A**, **B**, & **C** in terms of R, g, and h.
b) Determine the minimum height h that allows the block to slide through the vertical loop without losing contact with the road. Hint: consider an FBD at point **C** and think about the "losing contact" condition.

Note: if the block has non-negligible size, one must correctly reference the center of mass height at each position. This is annoying but not much different in terms of algebra…

8.5 A roller coaster engineer is considering three possible track designs shown below. Each of the three track designs has the riders start at the same height with negligible speed and go through a vertical loop. In each of the designs, the top of the loop is distance d above the ground.

Design A

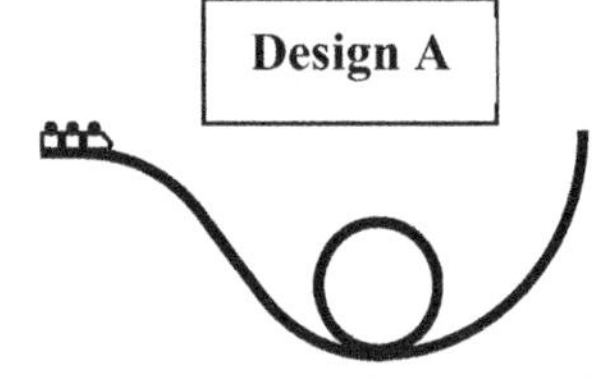

Design B

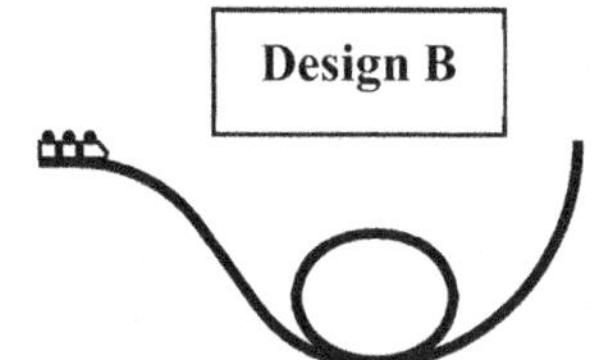

Design C

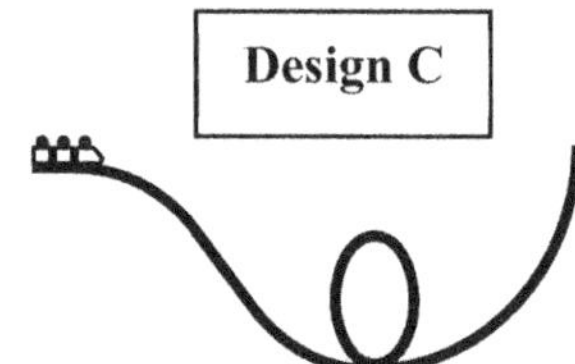

a) Rank the speeds for each design at the top of the loop.
b) Rank the accelerations for each design at the top of the loop.
c) Rank the normal force on the riders for each design at the top of the loop.

Reducing the radius of curvature reduces the minimum speed to remain in contact. This means lower initial height (and cost) is required. Tighter turns increase the accelerations felt by the rider. Tighter turns make car-to-car-couplings more challenging as well. I'm sure there is a lot more to think about but this sprang to mind…

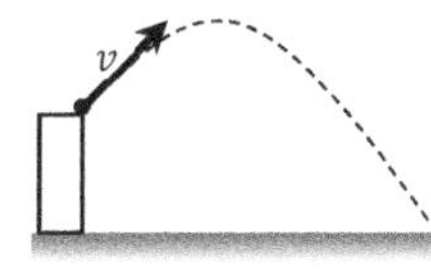

8.6 Feel free to skip projectile is shot from a cliff of height h with initial speed v at an unknown launch angle. Determine the impact speed of the projectile.

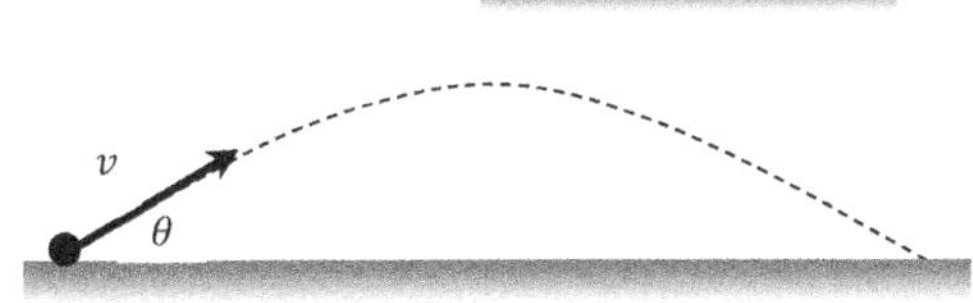

8.7 Feel free to skip, good review A 2 kg ball is launched with an initial speed of 40 m/s at an angle of 30°. The impact location is at the same height as the launch location. To make the math cleaner, assume $g = 10\frac{\text{m}}{\text{s}^2}$. Assume a reference level is chosen such that the intial gravitational energy is zero.

a) Before doing any calculations, sketch U_G vs t, K vs t, and E_{total} vs t. Think about when the plots should be zero or maximum and work from there. If you are completely clueless, move on to the next step.
b) Use kinematics to equations for speed and height versus time.
c) Use the above results to determine equations for $U_G(t)$, $K(t)$, and $E_{total}(t)$. Make plots with Excel or Matlab. Use 0.5 s increments for 4 s.
d) Think how would the plots change if you launched vertically instead?

DERIVATION OF SPRING POTENTIAL ENERGY ON PAGE 20. For now use $U_{spring} = \frac{1}{2}kx^2$ where k is a constant with units of $\frac{\text{N}}{\text{m}}$ and x is the distance of compression (or stretch) of the spring.

8.8 & 8.9 Consider a vertical mass spring simulation. One is discussed in the supplemental handouts.

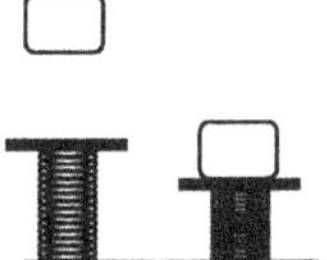

8.10 A block of mass m is initially held a distance h above a light spring platform. The mass is released from rest and lands squarely on the light platform. The platform acts essentially like a massless spring.

a) When the spring reaches max compression, which of the following quantities are zero: velocity, acceleration, net force?
b) Assume you know the spring compresses a maximum amount x. Determine the spring constant in terms of x, m, g, and h.
c) **Challenge:** assume you instead know k, m, g, and h. Determine the maximum compression.

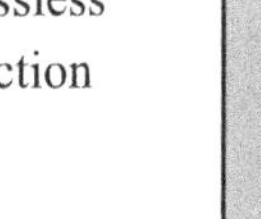

8.11 A mass-spring system is designed as shown. The spring, string and pulley are massless compared to the block of mass m. The pulley axle and incline both exert negligible friction forces. The spring has constant k and is initially unstretched.

a) What is the *equilibrium* position of the system?
b) What is the speed of the block when it first reaches the equilibrium position?
c) Which quantities are zero *at max stretch*: speed, acceleration, spring force, net force?
d) What is the maximum extension of the spring? Assume it never stretches far enough to touch the pulley.
e) Determine the magnitude of the net force on the mass at max extension in terms of m, g, and θ.
f) Make a bar graph showing relative sizes of U_G, U_S, and K for initial, equilibrium, and max stretch positions.

8.11½ A block (negligible dimensions, mass m) initially compresses a light spring distance x. Note: the block s not attached to the spring…just pressed against it. The block is released from rest and slides up a ramp of length d angled θ above the horizontal. On the ramp the block experiences frictional coefficients of μ_s& μ_k. The block launches from the top of the ramp traveling at speed v. Determine the spring constant used. Answer in terms of the given variables and the constant g.

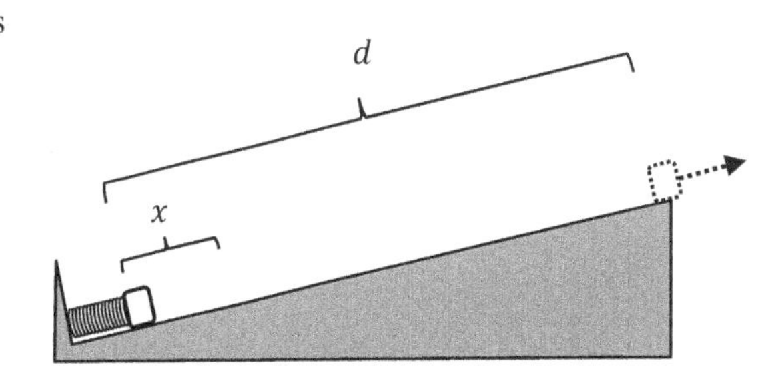

8.12 Feel free to skip…but it is pretty cool A 200 gram mass is used in a vertical mass-spring system. The spring constant is $k = 20\frac{\text{N}}{\text{m}}$. To make the math cleaner, assume $g = 10\frac{\text{m}}{\text{s}^2}$. The system is released from rest when the spring is in the unstretched position. If we consider the bottom of the motion as our reference level the vertical position of the spring versus time is given by

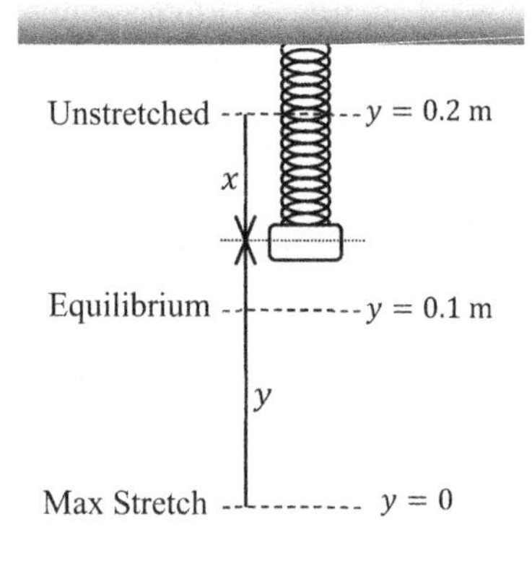

$$y(t) = (0.1\text{ m}) \times \left\{1 + \cos\left(10\frac{\text{rad}}{\text{s}}t\right)\right\}$$

a) Before using any numbers, sketch plots of U_G vs t, K vs t, U_s vs t, and E_{total} vs t over the course of one full oscillation.

b) Use the numbers to make plots in Excel or Matlab. Use 0.01 s increments for total time 0.63 s. Excel uses radians but so does our function…

8.13 A block of mass m is given an initial speed v on a level horizontal surface. It slows down and stops due to friction. The distance traveled while stopping is d. Determine the coefficient of friction.

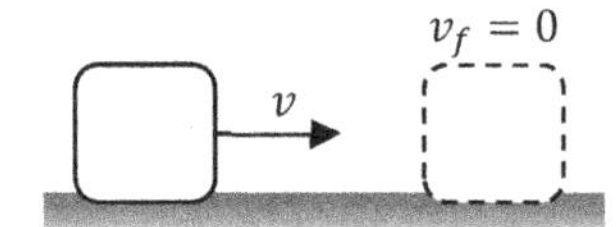

8.14 A block of mass m is pushed by a zombie across a level floor (thick arrow in figure). The applied force is angled θ below the horizontal. After traveling distance d, the block comes to rest. The coefficients of friction between the block and the floor are μ_s and μ_k.

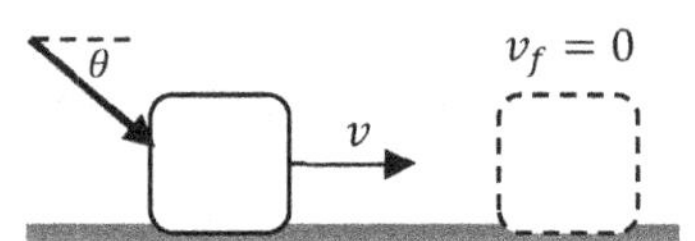

a) Determine the magnitude of the applied force.

b) Why do these zombies keep pushing all these boxes around? Don't they have brains to eat?

8.15 Consider a simulation. One is discussed in the supplemental handouts.

8.16 A simple pendulum is basically a mass at the end of a string of negligible weight. The size of the mass is much smaller than the length of the string. The radius of the simple pendulum is from the point where the string attaches to the pivot to the center of mass of the ball. A simple pendulum of length L is released from rest initially angled θ from the vertical.

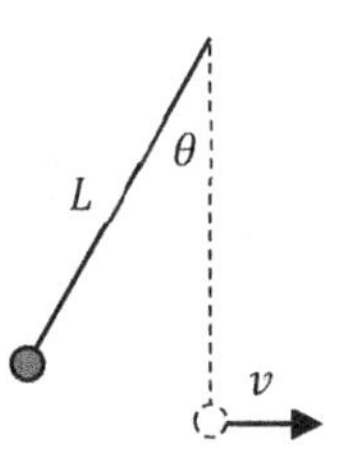

a) Determine the speed of the ball as it passes through the lowest point in the swing.

b) Determine the tension is the string as the ball passes through the low point.

8.17 A simple pendulum of length L is released from rest initially angled 90° from the vertical. At what angle will the ball reach half of max speed?

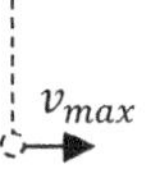

8.18 A Fletcher's trolley system is shown at right. The system is released from rest and accelerates downwards. Masses m & $2m$ are known.

a) When $\mu_k = 0.3$ the block reaches the ground with impact speed v. How high above the ground was the hanging mass when the system was released? Answer in terms of v and g.

b) Determine an upper limit on the coefficient of static friction.

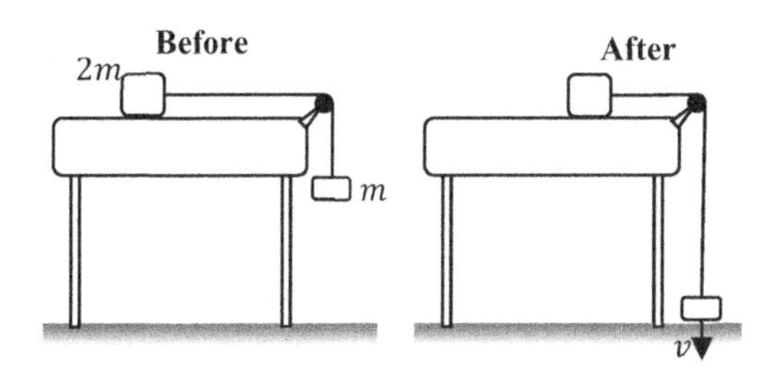

8.19 Consider the figure at right. Assume the incline is frictionless. The block on the incline has mass m while the hanging block has mass $2m$. Initially the hanging mass is held distance h above the ground. The system is released from rest. Determine the impact speed of the hanging block in terms of g, h, and θ.

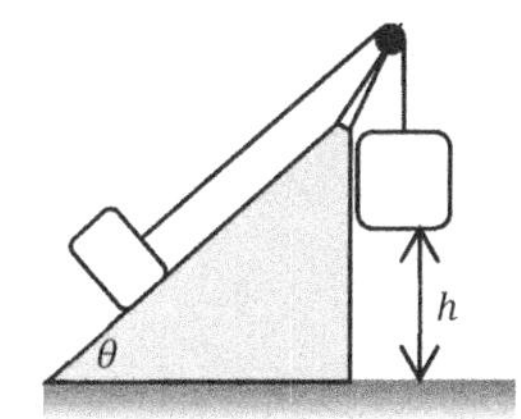

8.20 A ball of mass m swings in a vertical circle of radius r. Instead of being attached to a string, the ball is attached to a stiff metal rod of negligible mass is attached to frictionless pivot at one end.

a) What is the minimum speed of the ball required at the top of the loop for the ball to complete the circular motion? Why is the answer different for a stiff rod versus a string?

b) Assume the ball is released from rest at the top of the loop. It is given a very slight push to get the motion started clockwise. The push is so slight we may still assume the speed of the ball at the top is essentially zero. Determine the tension in the rod at the bottom of the swing terms of m, g, and r.

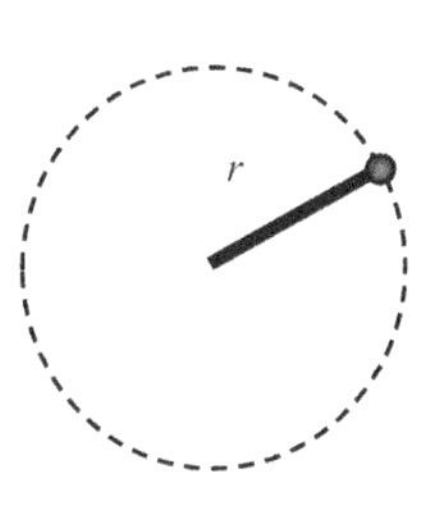

Summary comments/practical tips for doing energy problems:

- Before assuming conservation of energy is valid, think about possible sources of $W_{non-con}$ or W_{ext}
- Forces pointing towards the center during circular motion do no work. They change the *direction* of velocity but not the *speed*!
- In general, forces perpendicular to displacement do no work.
- For a two block system, give each block its own reference level.
- Internal forces do no net work on a two block system.
- On inclines the change in height is *usually* $h = L \sin\theta$.
- If a spring changes length the forces are **not** constant; use $E_i + W_{non-con} = E_f$.
- For pendulums remember how to use the picture below.

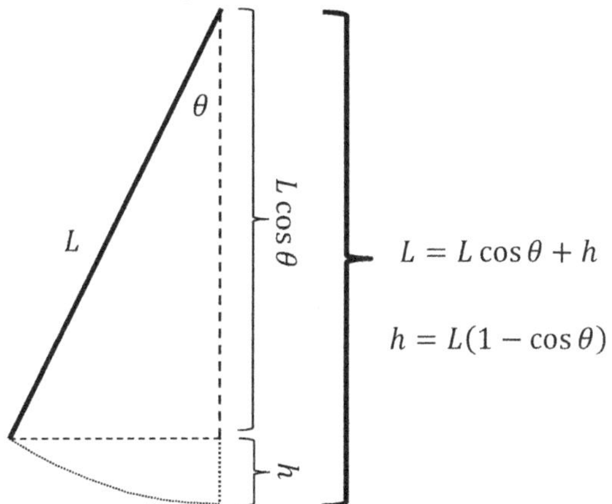

Using a Conservative Force to Determine Potential Energy
By definition

$$\Delta U = -W = -\int_i^f \vec{F} \cdot d\vec{s}$$

This assumes the force $\vec{F}$ is conservative (more on what this means next page). Solving this equation for U_f gives

$$U_f = U_i - \int_i^f \vec{F} \cdot d\vec{s}$$

For forces in the x-direction $d\vec{s} = dx\hat{\imath}$ and $\vec{F} = F_x\hat{\imath}$. The above equation simplifies to

$$U_f = U_i - \int_i^f F_x \, dx$$

If the initial position is zero we may rename $U_f = U(x)$ and get

$$U(x) = U_0 - \int_0^f F_x \, dx$$

Example: For a horizontal spring with unstretched position at the origin we know

$$\vec{F} = -kx\hat{\imath}$$

Notice, according to the above derivation, we must use

$$F_x = -kx$$

Here F_x is not the magnitude as usual. Annoying, yes, but hopefully you are somewhat used to this by having to constantly think about velocity vs speed.

$$U(x) = U_0 - \int_0^f F_x \, dx$$

$$U(x) = U_0 - \int_0^f (-kx) \, dx$$

$$U(x) = U_0 + \frac{1}{2}kx^2$$

If we assume the potential energy at the origin is zero we get

$$U_S(x) = \frac{1}{2}kx^2$$

This technique is used again in universal gravitation (Chapter 13) and electricity. Please learn it.

8.21 Suppose a horizontal mass spring system has spring constant $k = 100\ \frac{\mathrm{N}}{\mathrm{m}}$. Fill in the table for F_x and $U(x)$ below. Use those tables to sketch plots of F_x vs x and $U(x)$ vs x. **Note the units!!!**

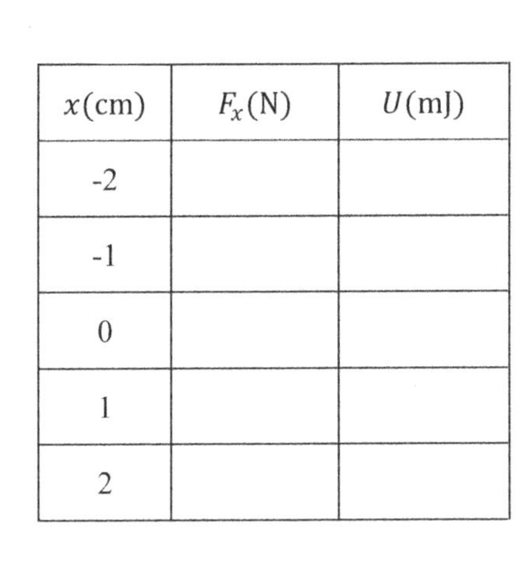

x(cm)	F_x(N)	U(mJ)
-2		
-1		
0		
1		
2		

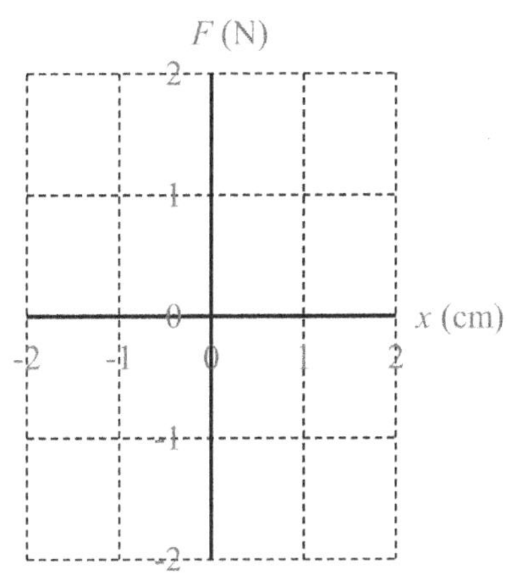

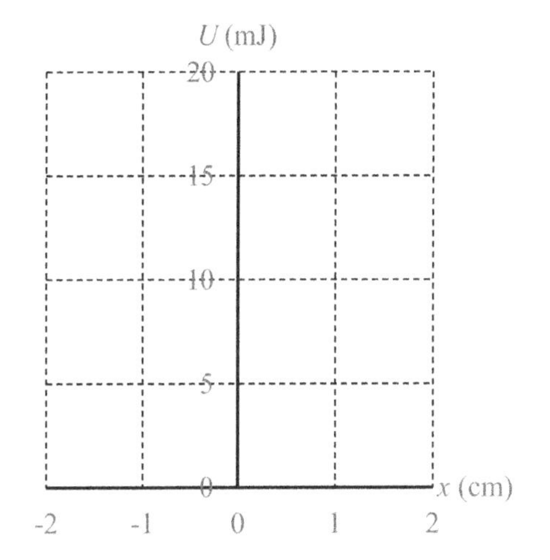

8.22 In chemistry and physics, matter is often modeled as if adjacent particles are connected by springs. In some situations, this model starts to deviate significantly from reality. Suppose the force between particles is instead

$$F_x = -kx + bx^3$$

where b & k are positive constants. Here x is the displacement of the right atom from the left atom. Assume $U = 0$ when $x = 0$.

a) Determine potential energy stored in the atoms as a function of position.
b) Determine the positions for which $U = 0$.
c) Determine the equilibrium positions.
d) Suppose $k = 10\ \frac{\mathrm{N}}{\mathrm{m}}$ and $b = 5000\ \frac{\mathrm{N}}{\mathrm{m}^3}$.

After class, plot F_x vs x and $U(x)$ vs x for values of x between -8 & $+8$ cm.

8.23 A block is dragged up a plane by a zombie at constant speed. The coefficients of friction between the plane and the block are both approximately μ. At the top of the plane the zombie slowly walks backwards and lowers the block to its initial position. During the lowering stage the zombie keeps the string taut with constant tension. Since this is a 1968-era zombie, the zombie walks so slowly the small acceleration stage as the zombie reverses direction is completely negligible (no post-2000 high-speed zombie issues here).

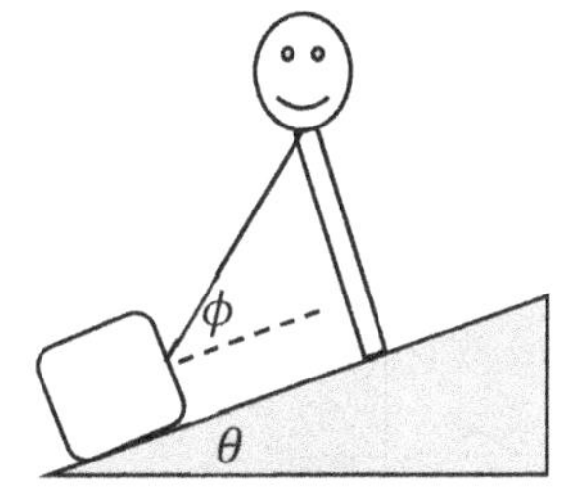

a) After class show $T = mg\frac{\sin\theta+\mu\cos\theta}{\cos\phi+\mu\sin\phi}$ on the way up and $T = mg\frac{\sin\theta-\mu\cos\theta}{\cos\phi-\mu\sin\phi}$.
b) Which force(s) in this problem do positive work on the way up then negative work on the way down?
c) Which force(s) do negative work then positive work?
d) Do any forces do only positive or only negative work?
e) Which forces are conservative? For these forces only you could use potential energy *instead* of work.

Note: a conservative force must satisfy the following criteria:

- Work done is path *independent.*
  This means "the same amount of work is done no matter how you travel from *any* point A to *any* point B."
- Work done on a *closed* path is zero.
  This means no *net* work is done going from *any* point A back to the same point A.

Using Potential Energy to Determine Force

To determine force from potential energy we can flip the definition.

$$F_x = -\frac{d}{dx}U(x)$$

Here are the practical things you will want to know:

- *Notice the minus sign.* Notice this implies that $\boldsymbol{F_x}$ **is opposite the slope of** $\boldsymbol{U(x)}$.
- When slope is zero, $F_x = 0$. The system is in equilibrium.
- Equilibrium points are classified as stable, unstable, or neutral depending on the concavity of $U(x)$. Stable equilibrium points are concave up, unstable are concave down, flat portions are neutral. See plot below.
- Think about how a marble would roll on the potential energy graph to get a feel for the direction and size of force. As an example, consider a marble released from rest at $x = -5$ nm on the plot below. We expect the marble to roll quickly to the right. This implies there is a large force to the right (large positive F_x).
- Units are often nm $= 10^{-9}$ m and meV $= 10^{-3}$ eV $= 1.6 \times 10^{-22}$ J. Note: 1 eV $= 1.6 \times 10^{-19}$ J.
- In many real-life situations it is much easier to create models for the energy (as opposed to forces) in a system. Once the energy is correctly modeled, forces can be determined with derivatives. This technique is used in quantum chemistry, quantum mechanics, and electrical engineering.

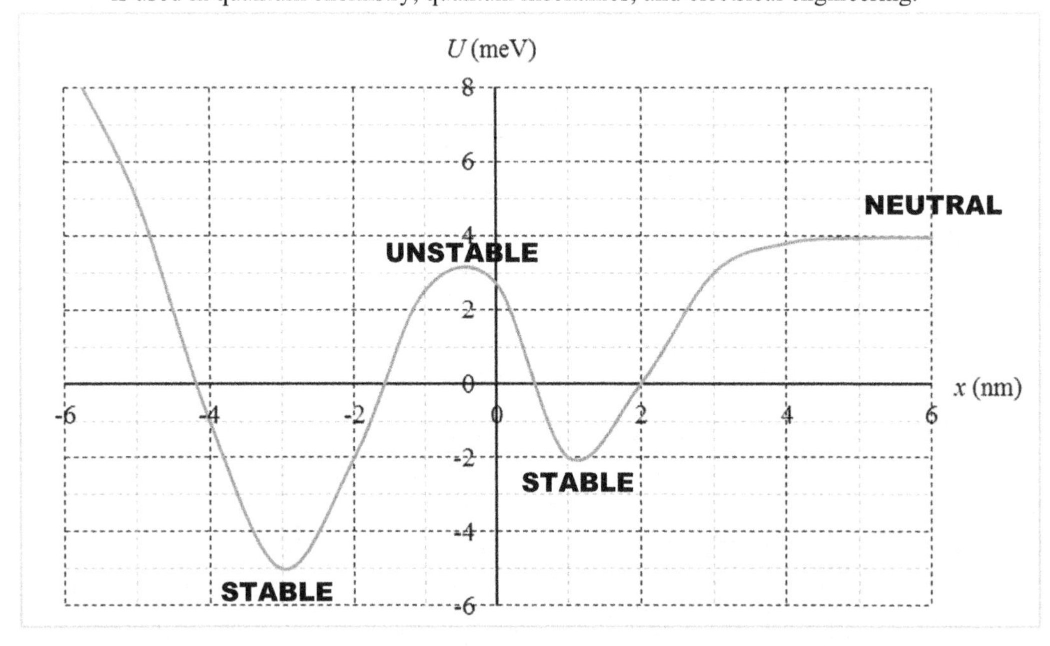

8.24 A potential energy function called the Lennard-Jones Potential is given by

$$U(x) = 4\varepsilon\left(\left(\frac{\sigma}{x}\right)^{12} - \left(\frac{\sigma}{x}\right)^{6}\right)$$

This potential energy models the interaction between two atoms. The 12th power term relates to Pauli repulsion while the 6th power term relates to an attractive van der Waals force. Assume σ and ε are positive constants and x is horizontal position. Units are discussed in part a.

a) What are the units on σ and ε? **For the rest of the page I will leave off units to reduce clutter. See note at the bottom of the page.**
b) Determine the force associated with this potential energy. This is the force one atom exerts on the other.
c) Determine the equilibrium position in terms of σ.
d) Which graph below represents F_x vs x and which represents U vs x? The plots use $\varepsilon = 1$ and $\sigma = 1$.
e) Compare the slope of U vs x to the values of F_x on the F_x vs x plot. Think about the size and sign of the force as well as the character of the equilibrium (stable, unstable, neutral).

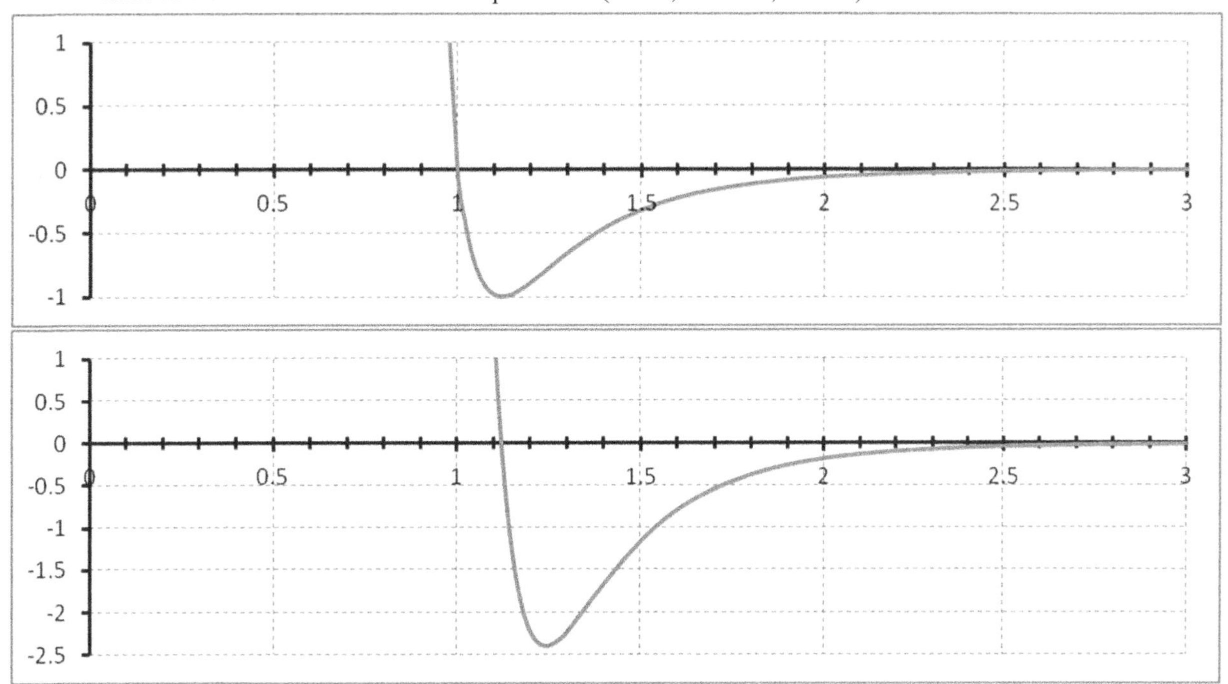

8.25 Now think about placing a marble on the potential energy plot (upper plot) at various locations.

a) If you placed a marble on the U vs x plot at $x = 1$, would it roll right or left? Compare your answer to the *sign* of F_x at $x = 1$.
b) If you placed a marble on the U vs x plot at $x = 2$, which way would it roll? Compare to F_x at $x = 2$.
c) Consider the marbles at $x = 1$ and $x = 2$ released from rest. Which would start rolling faster? Compare this to the *size* of F_x (ignoring direction) at $x = 1$ and $x = 2$.
d) Now consider $x = 1.12$. This corresponds to the minimum of $U(x)$ in the upper plot and the equilibrium point ($F_x = 0$) in the lower plot. A marble released from rest at this point on the graph wouldn't move at all! This is the point where Pauli repulsion and van der Waals attraction balance out and cause no net force on neighboring atoms. Any closer together and they get pushed apart. Any farther apart and they get pulled back together. Uhhh…I guess there is no question here, just a deep thought.

For molecular problems, typical units for x & σ are in nm, ε in milli-eV = meV (about 10^{-22} J).

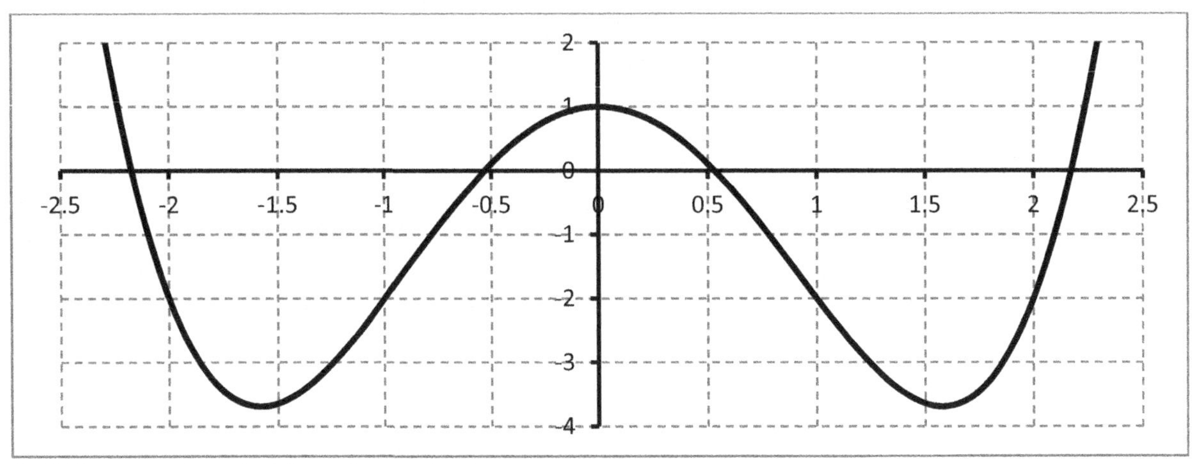

8.26 Now consider a double well potential given by

$$U(x) = 0.750x^4 - 3.75x^2 + 1.000$$

A plot of U vs x is shown above. **The units are x in nm and U in eV.** Note: 1 eV = 1.602×10^{-19} J.

a) Determine all equilibrium points. Identify the points as stable, neutral, or unstable equilibrium points.

b) Determine an equation for the force as a function of x associated with this potential energy function.

c) Determine the force at each point in the table at right. Negative numbers imply force acts to the left.

d) What are the units of this force? How do these units compare to N?

e) Sketch a plot of the force on the graph provided below.

f) Consider an electron initially located at the bottom of the left potential energy well (at the left minima). What minimum initial speed is required for an electron to travel to the right well?

x	F_x
-2	
-1	
0	
1	
2	

This type of potential might model an electron shared between two atoms in a covalent bond. The nuclei of the atoms would be located at the minima in the U vs x plot.

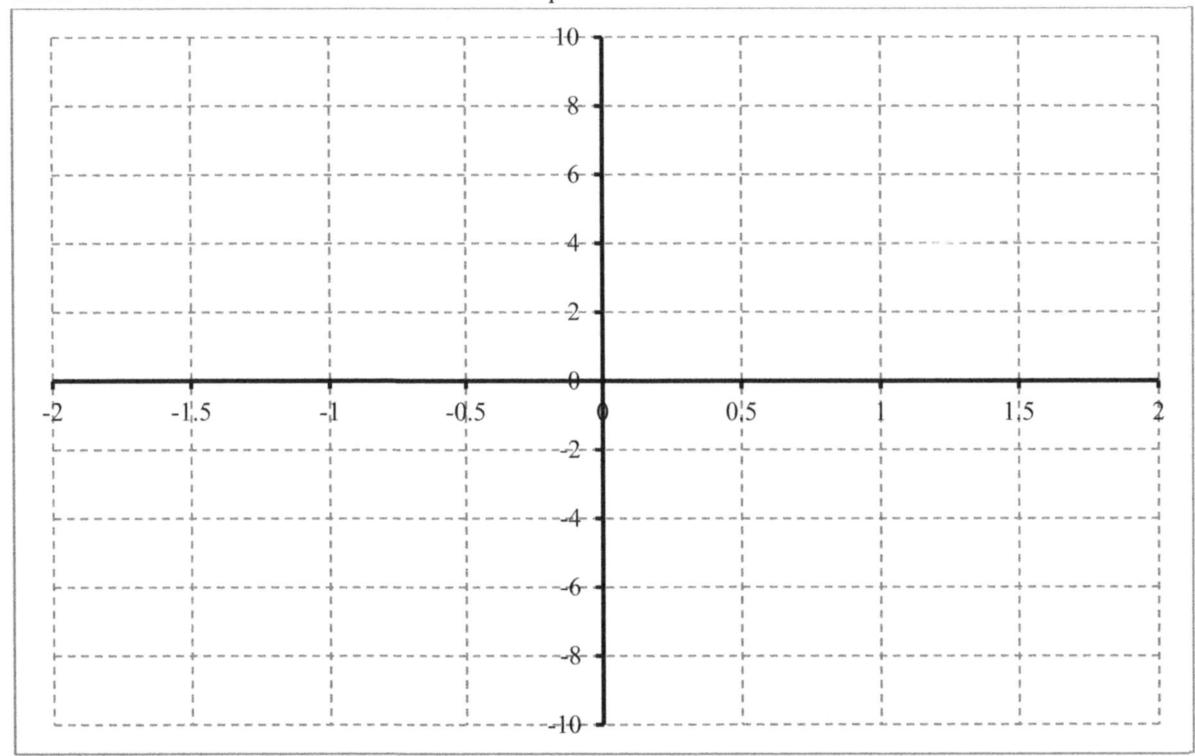

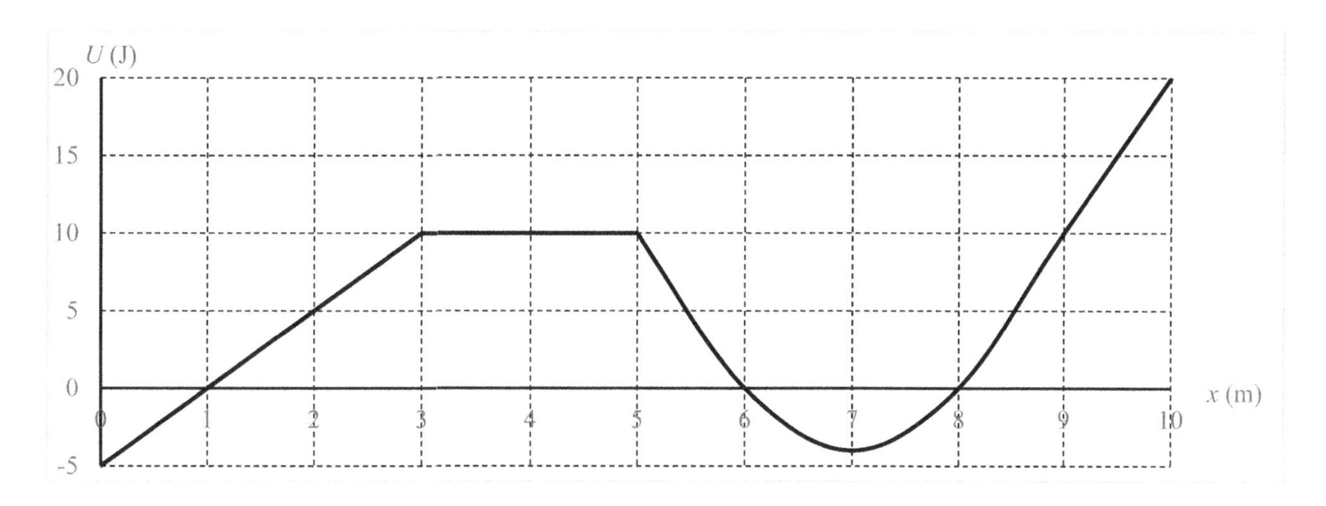

8.27 A plot of U vs x is shown above. Assume it shows a plot of potential energy associated with a conservative net force acting on a particle undergoing 1D motion along the x-axis.

a) Estimate the force acting on the particle at $x = 8.5$ m.
b) For what locations will the force act to the right on the particle?
c) For what locations will the force act to the left on the particle?
d) List all equilibrium locations and state if the equilibrium is stable, unstable, or neutral.

8.28 A 400 gram mass undergoes 1D motion along the x-axis. The net force on the mass is conservative. A plot of U vs x is shown above for the mass. The mass is initially at the origin moving to the right with total energy 15 J.

a) How far will the mass travel to the right before turning around? Explain using kinetic, potential, and total energy.
b) Determine the initial *speed* of the mass.
c) Determine the *velocity* of the mass when it first reaches $x = 5.5$ m.
d) Suppose the mass is released from rest at $x = 8.9$ m. Describe the motion of the particle after release.
e) Suppose the mass is released from rest at $x = 9.1$ m. Describe the motion of the particle after release.

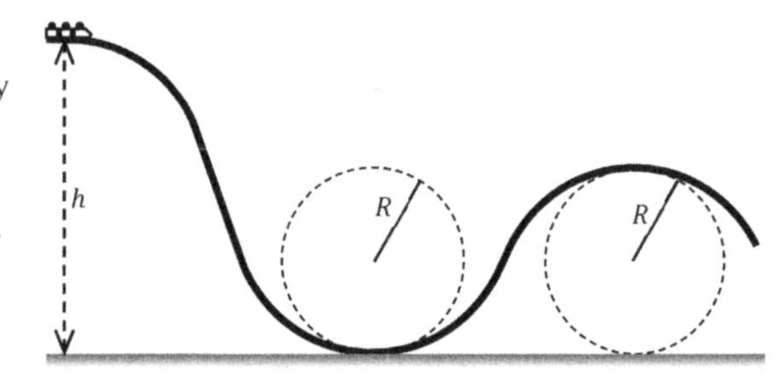

8.29 A roller coaster ride has negligible friction. The car begins *from rest* distance h above the ground and passes through a valley then over a hill with radius of curvature R. As each rider passes over the top of the lower hill the seat exerts a normal force equal to ½ of her or his weight. Don't worry; not seen in the figure is a giant pile of used physics scratch paper to cushion the impact of the car as it flies of the track at the right end.

a) Determine the initial height of the ride.
b) Determine the normal force on each rider (as a fraction of their weight) at the bottom of the valley.
c) By what maximum factor can the height be increased if the riders are still to remain in contact with the seat as they pass over the hill.

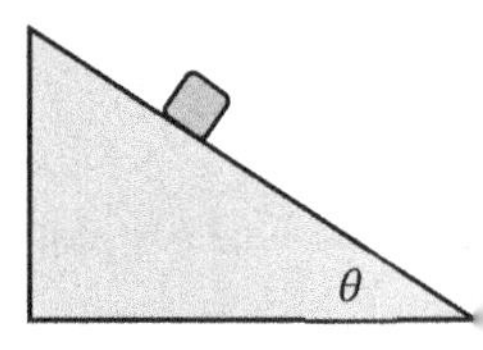

8.30 A 2.00 kg mass slides down a 30° incline for a distance of 4.8 m. The coefficents of friction are $\mu_s = 0.5$ and $\mu_k = 0.3$. Notice the critical angle is $\theta_c = \tan^{-1}(0.5) \approx 27°$. Assume all of the work done by friction is absorbed by the molecules of the ramp and block. The work done by friction increases microscopic motion (*internal* or *thermal* energy) of the molecules. We say $|W_f| = \Delta E_{int}$. To make the math cleaner, assume $g = 10\frac{\text{m}}{\text{s}^2}$.

a) Before using any numbers, sketch plots of U_G vs t, K vs t, ΔE_{int} vs t, and E_{total} vs t. Assume the bottom of the incline has $U_G = 0$.
b) Now use the numbers to determine equations and make plots. Use 0.25 s increments for a total time of 2 s.

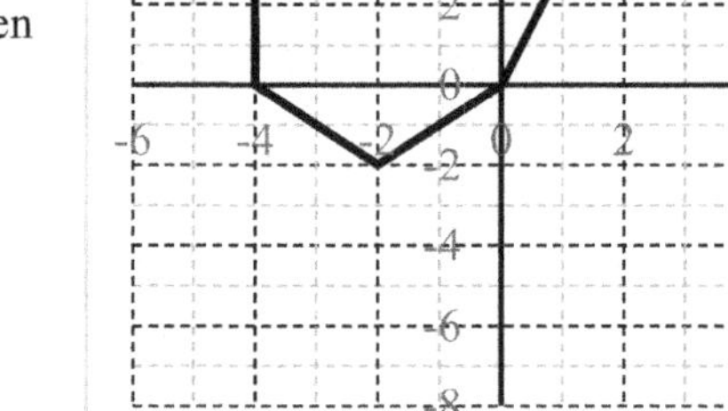

8.31 A plot of F_x vs x is shown at right for a conservative net force acting on a 2.00 kg particle. The potential energy reference level is chosen by saying the potential energy at the origin is zero.

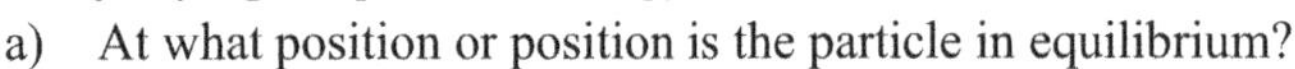

a) At what position or position is the particle in equilibrium?
b) A particle starts from rest at $x_i = -6.00$ m. If the particle does reaches the point $x = -2.00$ m, what is the speed when it reaches that point? If not, twiddle your thumbs…
c) How far to the right *will* the particle travel? Will it immediately move to the left? Will it reach the origin? Something in between? Something else? Does it oscillate back and forth?
d) What is the initial potential energy?
e) Plot U vs x.

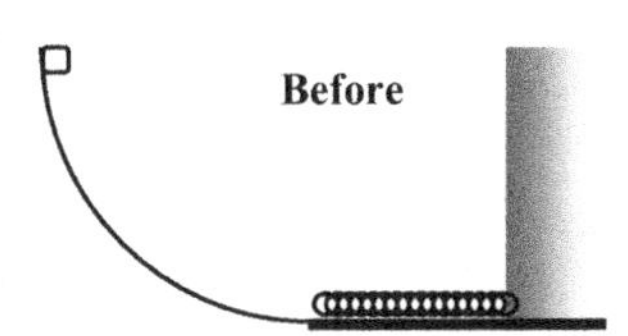

8.32 A block of mass m is released from rest at the top of a frictionless, curved ramp. The ramp is shaped like a quarter-circle of radius R. As it reaches the bottom of the ramp it enters a flat horizontal section that has a coefficients of friction μ_s & μ_k. At the instant the block hits the flat track it also impacts a spring of constant k.

a) Determine the max compression of the spring in terms of the givens and g.
b) **Verify:** Depending on the values of μ_s & μ_k, it is possible the block is on the verge of slipping upon reaching max compression. If this is true $kx_{max} = \mu_s mg$. It can be shown this occurs when $\mu_s = \sqrt{\frac{2kR}{3mg}}$ **and** $R = \frac{8mg}{3k}$ resulting in $x_{max} = \sqrt{\frac{2mgR}{k}}$.

   We expect the block remains at rest after whenever $x_{max} < \sqrt{\frac{2mgR}{k}}$.

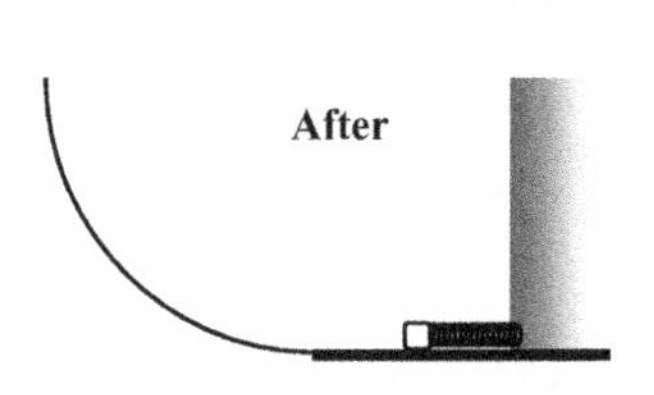

8.33 A block is released from rest and slides down an incline with friction. The block is initially distance L_0 from the spring. The portion of the incline below the spring has been polished to create a small region of negligible friction on the incline. Because of this, each time the block hits the spring it will rebound and head back up the incline with the same speed at which it impacts the spring. It reaches some new max position L_1. Assume incline angle and frictional coefficients are known.

a) Determine the difference between L_0 and L_1
b) Determine the ratio of L_1 to L_0.
c) Upon completing a second reflection, the block stops at max position L_2. Does block lose the same distance as in part a or reduce by the same factor as in part b?
d) In real life we know this can't go on forever. What determines if the block stops at max height (not touching the spring) versus at the bottom while touching the spring?

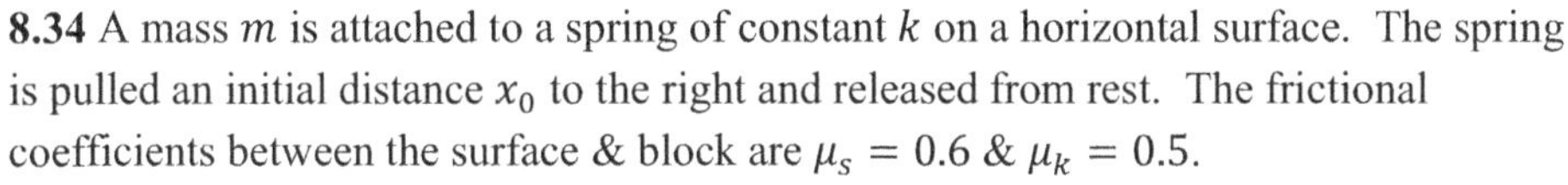

8.34 A mass m is attached to a spring of constant k on a horizontal surface. The spring is pulled an initial distance x_0 to the right and released from rest. The frictional coefficients between the surface & block are $\mu_s = 0.6$ & $\mu_k = 0.5$.

a) Determine the minimum value x_0 if the mass moves upon being released. Answer in terms of m, g, and k.
b) You are told $x_0 = 2.5\frac{mg}{k}$. Determine the speed v_1 when the mass first reaches the unstretched spring position.
c) Determine the distance x_1 the object travels to the left of the unstretched position during the first oscillation.
d) Where will the object finally come to rest? How many oscillations will the object undergo before coming to rest?

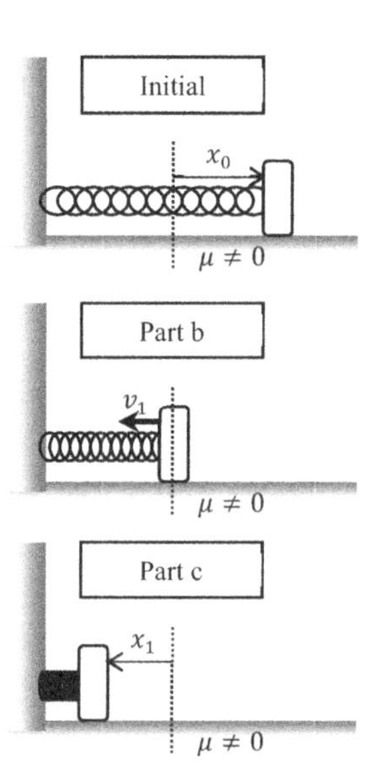

8.35 Not as tricky as it looks A rope of mass m and length L is initially at rest with half of its length off of a table. The rope is released from rest and slides of the table with negligible friction.

a) What is the speed of the rope at the instant it leaves the table?
b) Compare your results to the speed of a mass released from rest that falls distance $\frac{L}{2}$. Which should be faster? Explain.

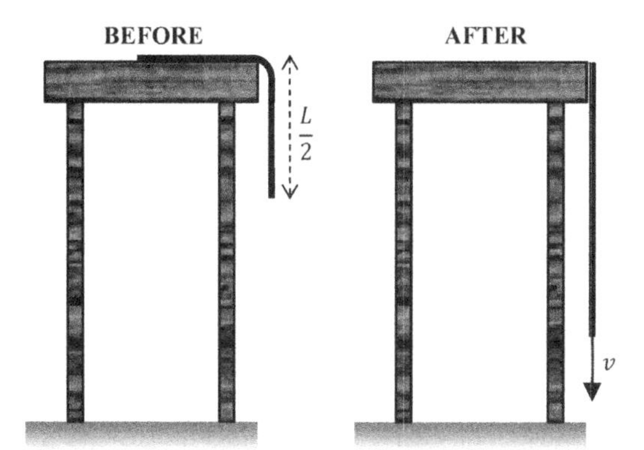

8.36 EXCELLENT PROBLEM A block of negligible size is at rest on the top of a frictionless, hemispherical dome of radius R. The block is given a very slight push to make it start sliding. Assume $v_i \approx 0$.

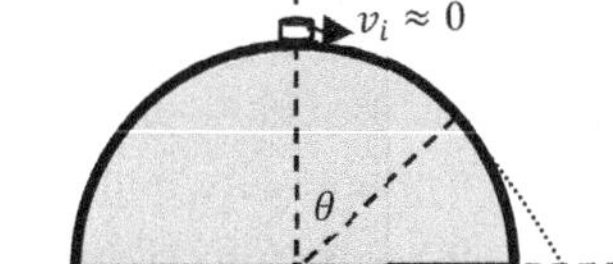

a) At what angle from the vertical will the block lose contact with the dome?

b) Assume $R = 1$. How far from the hemisphere does the block impact the earth?

8.37 A ball of mass m swings in a vertical circle of radius $r = 1.60$ m. The string can handle a maximum tension of $T_{max} = 50mg$ before breaking. At the bottom of the swing the string is on the verge of breaking. Friction is negligible. To keep the math clean, assume $g = 10\frac{\text{m}}{\text{s}^2}$.

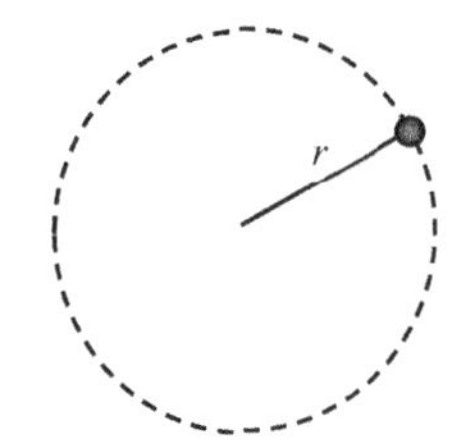

a) Determine the speed of the ball at the <u>bottom</u> in terms of r, g, and T_{max}.

b) Determine the speed of the ball at the <u>top</u> in terms of m, r, g, and T_{max}.

c) Determine the ratio of tension at the top to the tension at the bottom of the swing.

d) **Challenge:** Design a scenario where the string is on the verge of breaking at the bottom while at the minimum speed to complete the loop at the top? Is this possible? Explain.

8.38 A bungee jumper with mass m jumps off a bridge. The length of the bungee cord is L. To simplify the problem, assume the bungee cord is massless compared to the jumper. Also, assume air resistance is negligible to simplify the problem. Assume the jumper falls off the platform with negligible initial speed.

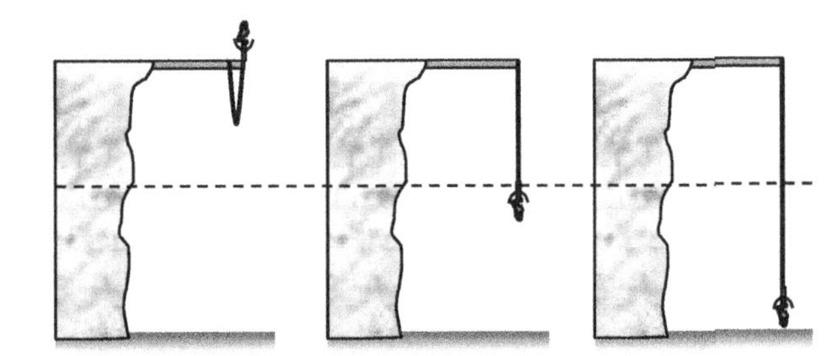

a) How fast is the jumper travelling at the instant the cord first begins to stretch? This point in time is indicated by the middle picture at right. This is usually called the fall to free length.

b) Assume the cord can be modeled as a spring with constant k. How far does the cord stretch?

c) How far below the platform is the jumper when she reverses directions at the bottom?

d) Determine a numerical result using $g = 10\frac{\text{m}}{\text{s}^2}$, $m = 60$ kg, $L = 15$ m, $k = 300\ \frac{\text{N}}{\text{m}}$. This approximately corresponds to a 130 lbs person using an approximately 50 ft cord.

e) Use your numbers to determine the jumper's acceleration at the bottom? Express in g's.

f) It is possible to jump in such a way that you are swinging like a pendulum as well as plummeting downwards. How would this affect the max stretch of the cord?

Now that you know the simplified model, you can probably understand what you read online. Try a web search for "bungee jump physics".

8.39 Challenge: A student hangs inverted from his toes from a platform distance h above the ground. In each hand he holds a mass m. One of the masses is connected with a chain of mass m to the platform. In contrast to a bungee jump, the chain has negligible stretch. The chain has total length $2.1h$ to allow both masses to reach the ground. The two masses are released simultaneously. Which hits the ground first? Is it a tie? Explain your answer.

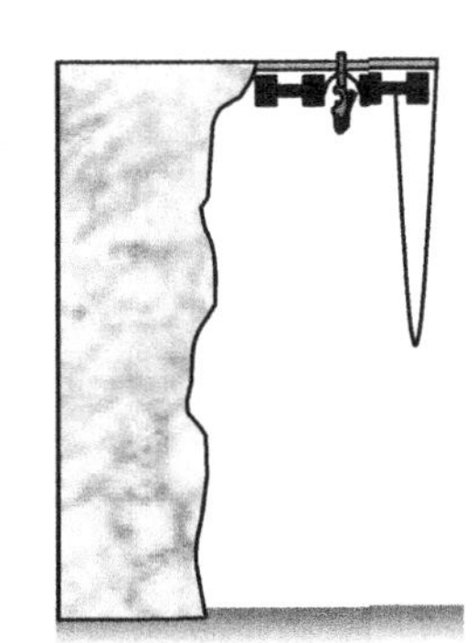

8.40a Explain 2 reasons why this force equation is <u>not valid</u>:

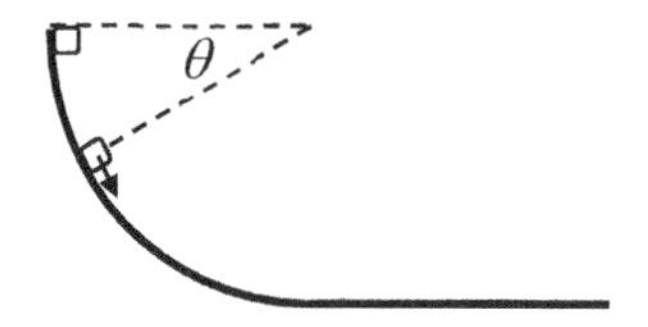

Consider a puck of mass *m* (negligible height) released from rest at the top of a quarter-circle track of radius *R*. There is friction with coefficient μ_k between the puck and the wall. A student does a free body diagram and determines, incorrectly, the frictional force is given by

$$f = \mu_k mg\cos\theta$$

8.40b Why is this still incorrect?

Another student fixes the errors then adds:

"The frictional force is given by

$$f = \mu_k m\left(g\,sin\,\theta + \frac{v^2}{R}\right)$$

Furthermore, I can quickly find the velocity using an energy problem. In general I know that for objects released from rest

$$v = \sqrt{2g|\Delta h|}$$

where $|\Delta h|$ is the distance fallen. For this diagram, I can see $|\Delta h| = R\,sin\,\theta$. Therefore

$$f = 3\mu_k mg\,sin\,\theta$$

I can then get the work done by friction using

$$W = \int_0^{\pi/2} \vec{f}\cdot d\vec{s}$$

$$W = \int_0^{\pi/2} -(3\mu_k mg\,sin\,\theta)(Rd\theta)$$

Here the minus sign accounts for the fact that friction is always directed opposite the displacement. On a circle the differential arclength is given by $ds = Rd\theta$. Finishing the calculation gives

$$W = -3\mu_k mgR\int_0^{\pi/2} sin\,\theta\,d\theta$$

$$W = 3\mu_k mgR\,cos\,\theta|_0^{\pi/2}$$

$$W = -3\mu_k mgR$$

I mean, like, ya know?"

What is wrong with this explanation?

8.40c Doesn't this type of problem seem like an ideal candidate for a coding exercise?

- Assume the origin of the coordinate system is located at the center of the circle. This allows us to use the constraint equation $R^2 = \sqrt{x^2 + y^2}$.
- Release the block from rest. Calculate the initial force at this position. Use this force to update momentum. Update to vertical position. Use the constraint to also update the horizontal position.
- Rotate the block by an appropriate angle based on the new position.
- Inside a while loop,
    - Use the updated angle to update the normal force and frictional forces.
    - Use the updated forces to update momentum.
    - Use updated momentum to update vertical position.
    - Use the constraint to update horizontal position.
    - Use the updated positions to compute the updated angle.
    - Increment the time and repeat.
- Add code which exits the loop after the block has rotated 90°. Code it slowing to a stop on level ground.

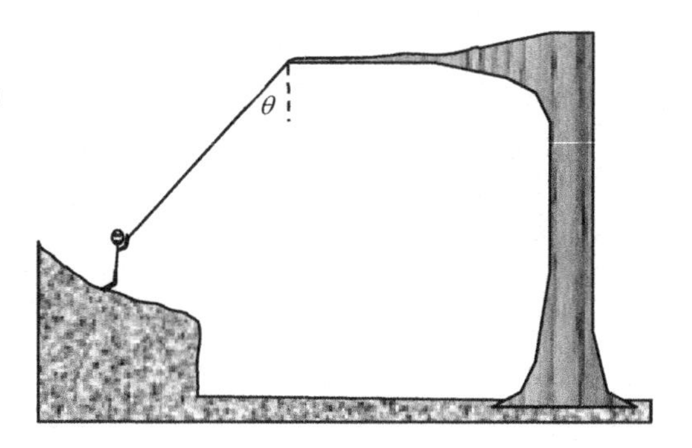

8.41 A certain vine hangs from a tree and can support a maximum load of 800N. The vine has unknown length. Jane swings on the rope by raising the vine to some angle and starting from rest as shown in the figure. Jane has weight 600 N. What is the largest starting angle Jane can use without breaking the rope?

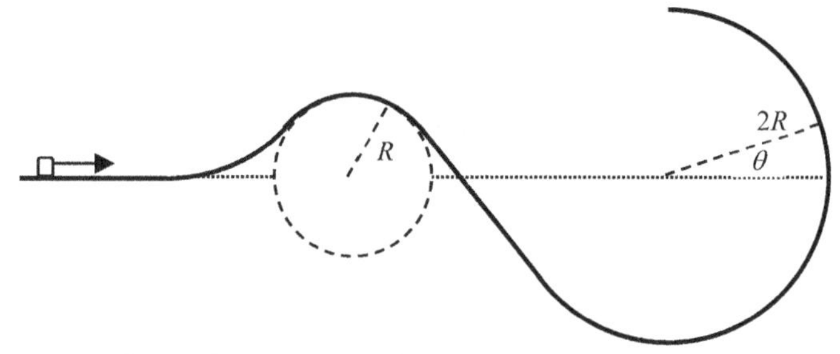

8.42 A block of negligible size slides with negligible friction on a track into the unusual shape shown. The hill has a radius of curvature exactly half that of the 2nd circular portion of the track. The block has initially moving at the maximum speed possible such that it just barely remains in contact with the road as it goes over the top of the hill.

a) What is the initial speed of the block?
b) At what angle will the block separate from the track? Will it make it to the end ($\theta = 90°$) or fall off before reaching the end ($\theta < 90°$)?
c) Suppose the second radius was $3R$ instead of $2R$. Does your previous answer change?

MOMENTUM

Up to now we have discussed a single object translating around (or a group of objects translating in unison).

Sometimes things collide (car crash, balls bouncing off walls, bullets embedding in ribs, etc).
Sometimes single object explodes in multiple parts (bombs or separating spacecraft).
Sometimes system of objects rearranges positions without external forces (zombie in boat walks to other end).
Momentum and center of mass help us understand these types of events.

Term	Symbol	Comment
Momentum	$\vec{p} = m\vec{v}$	Momentum is a vector (you have momentum *to the right*, or *left*, etc)
Collision Time	Δt	Usually I get lazy and shorten Δt to simply t This is the time two objects are in contact with each other
Average Force	$\vec{F}_{1on2} = \frac{\Delta \vec{p}_{of2}}{\Delta t}$	*Average* force exerted **BY** object 1 **ON** object 2 is *change* in momentum **OF** 2 divided by the time two objects are in contact
Impulse	$\vec{J} = \Delta\vec{p}$ $\vec{J} = \vec{F}\Delta t$	Typically we say impulse is *change* in momentum *of a single object* Useful when coding animations of motion by doing the following: • Calculate net force vector based on *current* position • Update the momentum of the system using $\vec{p}_f = \vec{p}_i + \vec{F}_{net}\Delta t$ • Compute current avg velocity using $\vec{v}_{avg} \approx \frac{\vec{p}_f}{m}$ • Update the position using $\vec{r}_f = \vec{r}_i + \vec{v}_{avg}\Delta t$

9.1 Suppose a 0.075 kg bouncy ball is moving 6.0 m/s to the left. It hits a massive vertical cliff. It bounces and moves 4.0 m/s to the right. The ball is in contact with the cliff for 0.050 sec. During the collision there *is* an external force (gravity). Fortunately, because the collision time is very short, we can ignore the effects of the external force and use conservation of momentum.

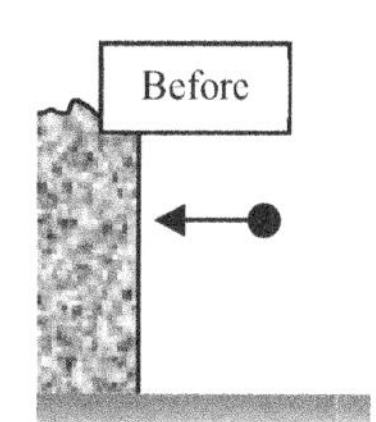

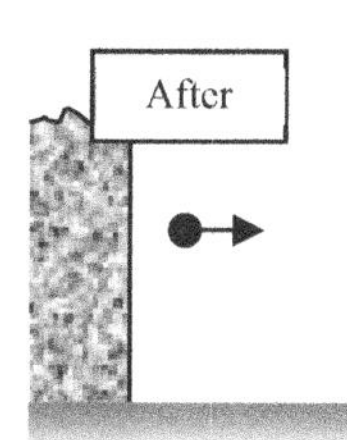

a) What is the *initial momentum* of ball?
b) What is the *final momentum* of ball?
c) What is the *change in the momentum* of ball?
d) What is another name for the change in momentum of ball?
e) What is the *average force* on ball during collision (magnitude and direction)?
f) What is the *average force* on earth (magnitude and direction)?
g) Determine the change in the earth's velocity as a result of the collision. Is it significant? The mass of the earth is approximately $M_E \approx 6 \times 10^{24}$ kg.

9.2 A 0.250 kg ball is dropped vertically from unknown initial height. The ball hits the ground and bounces back up. A strange alien chemically links to your mind to inform you the upwards impulse experienced by the ball during the collision with the ground is $10.0\ \mathrm{kg} \cdot \frac{\mathrm{m}}{\mathrm{s}}$. During the collision, the alien's telekinetic chemical link also tells us the ball loses 4.00 J of energy. Determine the speed of the ball just before and just after impact with the ground.

Note: **9.48** is an awesome review question tying together graphical interpretation of a xt-plot & impulse. I think it gives a more realistic idea of how one might use video capture to determine average force during impact.

Conservation of Momentum

$$\vec{F}_{2on1} = -\vec{F}_{1on2}$$

$$\frac{\Delta\vec{p}_{of1}}{\Delta t} = -\frac{\Delta\vec{p}_{of2}}{\Delta t}$$

$$\Delta\vec{p}_{of1} = -\Delta\vec{p}_{of2}$$

$$\Delta\vec{p}_{of1} + \Delta\vec{p}_{of2} = 0$$

$$\Delta\vec{p}_{1\,\&\,2\;system} = 0$$

$$\vec{p}_{\substack{system\\final}} - \vec{p}_{\substack{system\\initial}} = 0$$

$$\vec{p}_{\substack{system\\initial}} = \vec{p}_{\substack{system\\final}}$$

Things to notice:

- Conservation of momentum is a different way of thinking about Newton's 3rd law (action-reaction)
- Momentum of *single objects will change* during a collision
- Momentum of the entire *system* of objects *will not change* (is conserved)

Practical Conservation of Momentum Problem Solving Strategy:

1) First identify if it is appropriate to use conservation of momentum
    a. Is it a collision/explosion/change of shape with negligible external forces?
    b. If collision time is small (typically true) even large external forces like gravity are negligible.
2) Draw a before and after picture.
    a. If you are unsure of a direction pick one and go with it.
        i. For example, suppose you aren't sure of the final speed v_f.
        ii. If your final answer for v is negative, the final object moves opposite the direction drawn.
3) List your conservation of momentum equations in the *x* and *y* directions
    a. $m_1 v_{1ix} + m_2 v_{2ix} = m_1 v_{1fx} + m_2 v_{2fx}$. Similar for y-direction.
    b. WATCH OUT! Often $v_{1ix} = v_{1i}\cos\theta$. DON'T WRITE $v_{1ix}\cos\theta$. See the difference?
4) Read for keywords to see if kinetic energy is also conserved
    a. If ELASTIC then $E_i = E_f$. Usually this means $K_i = K_f$.
    b. This usually gives $m_1 v_{1ix}^2 + m_2 v_{2ix}^2 = m_1 v_{1fx}^2 + m_2 v_{2fx}^2$ because the ½'s all drop out.
5) Re-read the problem, plug in things that are zero, solve for something.

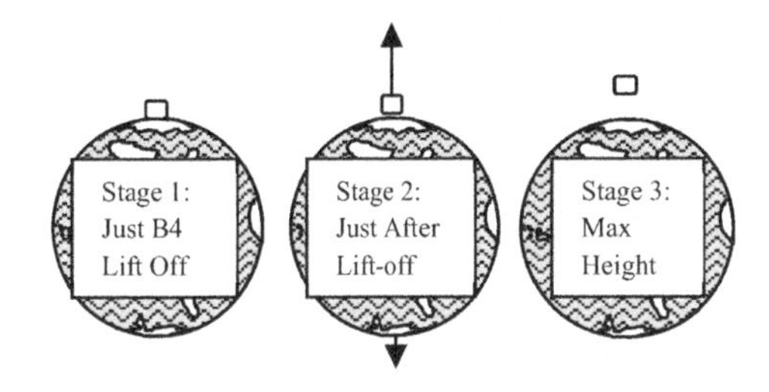

9.3 Suppose one day everyone on earth decides to gather at the north pole and jump at the same time. How much might this affect the earth? First we must make some assumptions. For now we can be bold and assume everyone is tightly grouped, has a large mass, and can jump as well as Muggsy Bogues in his prime. Muggsy was the shortest player in the NBA (5 ft 3) and had a vertical leap (>1 m).

a) Estimate the number of people on earth. If in doubt, use the largest number you think is realistic.
b) Estimate the average mass of everyone on earth. If in doubt, use the largest number you think is realistic.
c) Estimate the lift-off speed of everyone as they jump. Assume everyone on earth can raise her or his center of mass to a max height of 1 meter. Use energy to compare just after lift-off to max height.
d) Do a conservation of momentum problem comparing just before to just after lift-off. The people are tightly grouped so they act like a block. Figure is obviously not to scale. For the instant just after the jump, assume all people on earth are moving upwards with the speed you estimated in the previous part. The mass of the earth is approximately $M_E \approx 6 \times 10^{24}$ kg. Use this problem to determine the recoil speed of the earth.
e) Heyo! Did you notice I said recoil *speed* in the previous part? In momentum problems we must always read carefully to see if statements use speed or velocity. The two are related but not identical!

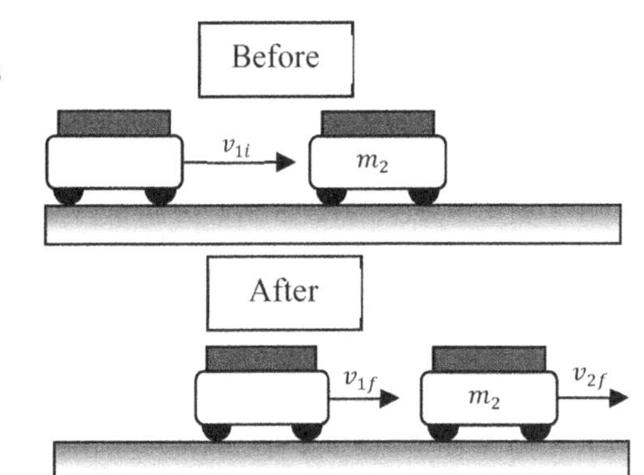

9.4 An object with mass m_1 moves to the right with initial *speed* v_{1i}. Mass m_2 is motionless. The two objects collide. After the collision the we observe m_2 moves away from m_1 with *speed* v_{2f}. Because I was so focused on the 2nd mass, I forgot to watch what happened to the 1st mass after the collision. Assume the *speed* of m_1 after the collision is unknown. For now assume both objects move right after the collision…

a) Use conservation of momentum equation to determine the final **speed** of m_1 in terms of the other variables.
b) Let $m_1 = 0.50$ kg, $m_2 = 1.00$ kg, $v_{1i} = 6.0\frac{\text{m}}{\text{s}}$, and $v_{2f} = 3.8\frac{\text{m}}{\text{s}}$. Determine the final **velocity** of m_1.
c) Determine the change energy during the collision. The energy in each stage (before or after) is simply the sum of kinetic energy for each object.
d) Where did the lost energy go?
e) Think: what external forces act on the objects during the collision? Why are we still able to do conservation of momentum even though external forces are present?

Comment: Notice in the figure these blocks are *rolling*. Because the wheels have low friction axles and low mass (compared to the carts) this problem can be modeled as *sliding* with no friction. We will handle rotation later…

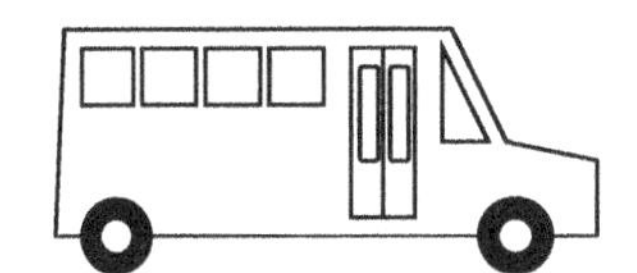

9.5 A school bus is driving to the right and collides with an insect flying to the left. During the collision the bug sticks to the windshield.

a) During the collision, who *exerts* the larger *force*?
b) During the collision, who experiences the larger *change in momentum*?
c) During the collision, which both objects (fly and bus) *accelerate*. Which acceleration has larger *magnitude*?

Types of collision problems

The names I use for collisions will not match up perfectly with textbooks. Most books only use three types (elastic, inelastic, and perfectly inelastic). I give the explosion type problems their own name just because it matches how I speak in class. Strictly speaking, there is no need for them to be considered separately.

Name	Comment
***Perfectly* Inelastic** Objects stick together $\vec{p}_i = \vec{p}_f$ $E_i \neq E_f$	After collision objects form a single mass moving together in unison. As a result, instead of having v_{1f} & $v_{2f}$the final speed of combined object is v_f Energy is *lost* ($\Delta K < 0$). Examples: two cars collide and get stuck together, a lump of clay hits (and sticks) to a block, problem **9.6**
Elastic Perfectly "bouncy" $\vec{p}_i = \vec{p}_f$ $E_i = E_f$	After collision objects do <u>not</u> form a single mass. Energy is *conserved* ($\Delta K = 0$). Examples: two pool balls collide and lose negligible energy, a steel ball bounces off a steel plate, springs
Inelastic *Imperfectly* "bouncy" $\vec{p}_i = \vec{p}_f$ $E_i \neq E_f$	Most real life collisions fall into this category. Many real life collisions can be approximated by one of the other categories. Energy is *lost* ($\Delta K < 0$) HOWEVER blocks do not form a single mass. Example: problems **9.1** & **9.4**
Typical Explosion $\vec{p}_i = \vec{p}_f$ $E_i \neq E_f$	Essentially a perfectly *inelastic* collision played in reverse. Energy is *gained* ($\Delta K > 0$). The gained energy is typically converted from chemical potential energy. Initially the bomb has speed v_i Afterwards the fragments have speeds v_{1f} & v_{2f} Examples: a bomb explodes and breaks into two pieces (chemical energy converted to kinetic to cause explosion), problem **9.3** (humans are required to convert chemical potential energy to jump)
***Elastic* Explosion** $\vec{p}_i = \vec{p}_f$ $E_i = E_f$	Variation of explosion where a conservative force is used to cause blocks to spread apart Energy is *conserved.* Examples: blocks pushed apart by massless springs, magnets are used to push blocks apart, block slides down a frictionless wedge. See problems **9.15** & **9.16**

9.6 A cart of mass m_1 slides with speed v towards a stationary cart of mass m_2. After the collision the two blocks slide together to the right. To achieve this result, someone put small pieces of Velcro on the blocks to ensure they stay together after the collision. Assume friction is negligible.

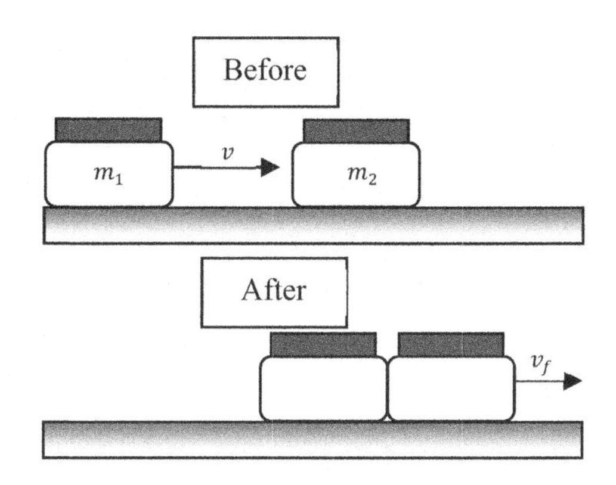

a) What type of collision is this?
b) Determine the final speed after the collision.
c) Determine the % change in energy. Percent change in energy is given by the formula $\%\Delta K = \frac{K_f - K_i}{K_i} \times 100\%$. Think: should your result be positive or negative?
d) Use your previous results to fill in the table below. The first line is done for you to show you what I am looking for.

m_1	m_2	v_f	$\%\Delta K$
m	m	$\frac{v}{2}$	-50%
$3m$	m		
m	$3m$		

Notice, for perfectly inelastic collisions, when a moving heavy object hits a light stationary object (2nd row of table) not much energy is lost. On the other hand, when a moving light object hits a heavy stationary object (3rd row) almost all of the energy is lost. Again, this reasoning is based on a perfectly inelastic collision.

9.7 A cart of mass m_1 slides with speed $v_{1i} = v$ towards a stationary cart of mass m_2. After the collision the blocks move with different speeds. Magnets were placed between the carts so they never actually touch each other during the collision!!! Assume friction is negligible.

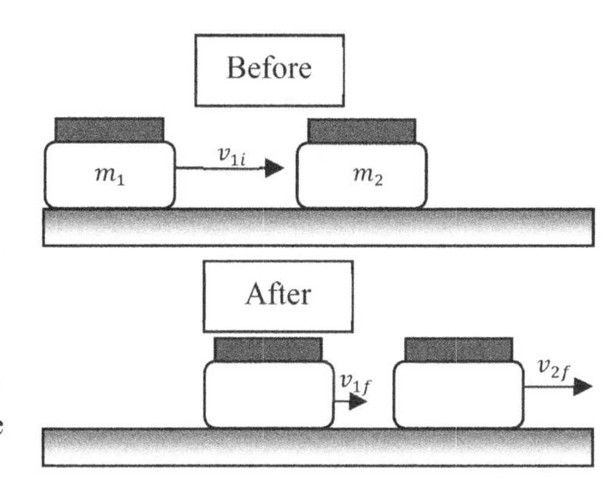

a) What type of collision is this?
b) Determine the final speeds after the collision. For now, assume my arrows are drawn the correct way.
c) Determine the % change in energy.
d) Use your previous results to fill in the table below. The second line is done for you to show you what I am looking for. Think about the meaning of the signs on your answers for the third row.

m_1	m_2	$\vec{v}_{1f}$	$\vec{v}_{2f}$
m	m		
$3m$	m	$\frac{1}{2}v\hat{\imath}$	$\frac{3}{2}v\hat{\imath}$
m	$3m$		

9.8 Variables renamed from older editions…A table has a hole cut out of it. A block of mass M is placed at rest over the hole. A bullet of mass m is shot with an initial velocity of v_1 at the center of a block of mass M (see top figure at right). The block is initially at rest. The bullet embeds in the block and rises to maximum height h. Assume m is so small it does not significantly change the center of mass position of the block after it embeds.

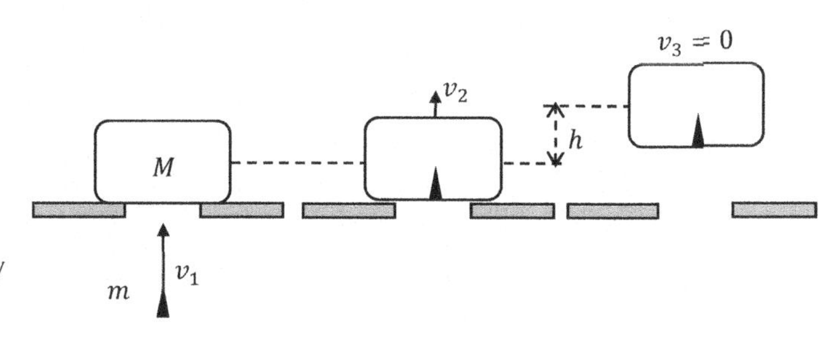

a) Show the max height is $h = \left(\frac{m}{M+m}\right)^2 \frac{v_1^2}{2g}$.

b) Show the energy change *during the collision* is

$$\%\Delta E = \left(\frac{m}{m+M} - 1\right) \times 100\% = \frac{-M}{m+M} \times 100\%$$

9.9 Class demo A ramp of length L is inclined at a small angle ($\theta < 8°$). A cart of mass m_c is released from rest. The cart can be modeled as sliding with negligible friction down the ramp. At the bottom of the ramp the cart slams into a sled of mass m_s on the flat ground. The cart sticks to the sled and the two slide off together until coming to rest in distance d. A set of four stages is drawn at right. Figures not to scale.

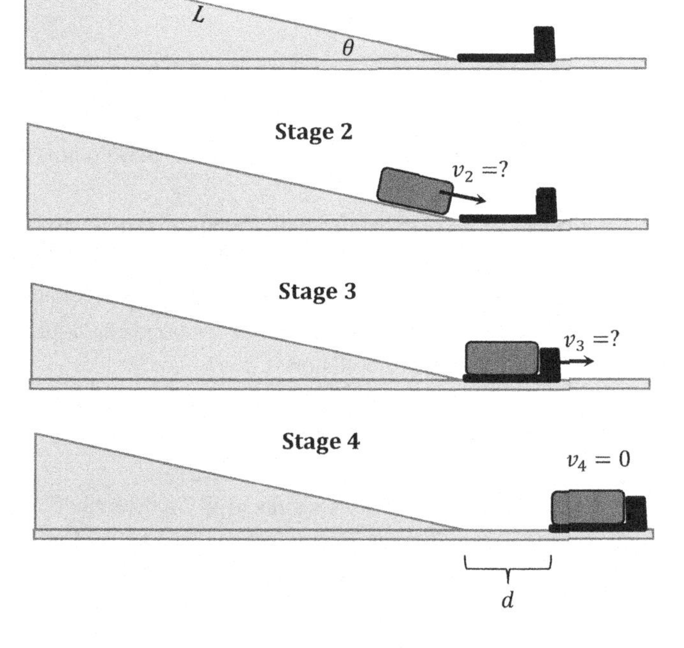

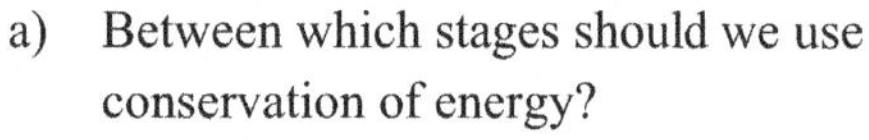

a) Between which stages should we use conservation of energy?

b) Between which stages should we use conservation of momentum?

c) Upon plugging in numbers, $\cos 8° \approx 0.99$. This means the horizontal velocity component at the end of the ramp is 99% of the cart's total speed. If we assume $v_{2x} \approx v_2$, determine the coefficient of friction between the ground and sled.

d) Just before the collision at the end of the ramp the cart has downwards vertical momentum. Just after the collision there is no discernible vertical momentum? I thought momentum was conserved in a collision…what gives?

e) The cart has a mass of about 500 g while the foam sled is closer to 30 g. Which object experiences greater force magnitude during the collision?

f) Which object experiences the larger acceleration magnitude during the collision?

g) Which object experiences a larger magnitude change in momentum during the collision?

9.10 Variables renamed from older editions…A hole is cut in a table. A block with mass $M = 2.00$ kg rests over the hole. Just before impact with the block, a bullet of mass $m = 10.0$ g moves upward with initial speed $v_{1m} = 1000\ \frac{\text{m}}{\text{s}}$. It strikes and passes through the block. The bullet emerges from the block with a speed of $v_{2m} = 400\ \frac{\text{m}}{\text{s}}$. In the process of passing through the block, the bullet bores out a small hole in the block. In real life the block will lose a tiny bit of mass and the bullet would be deformed. To simplify our model, assume the block is pretty hefty and the lost mass is negligible. Also assume the bullet is not deformed at all. After the collision the bullet flies far into the sky while the block rises to some maximum height. **For this problem draw three pictures:** just before the collision (done for you), just after the collision, and when block reaches max height.

a) Think: For which set of figures is conservation of momentum valid? For which set of figures is conservation of energy valid? Do we need to worry about the bullet, the block, or both when considering kinetic energy?
b) Algebraically determine the speed of the block just after the bullet has passed through it. Then check by plugging in numbers.
c) Determine the max height reached by the block. Do this entirely algebraically, do a unit check, and then plug in the numbers.

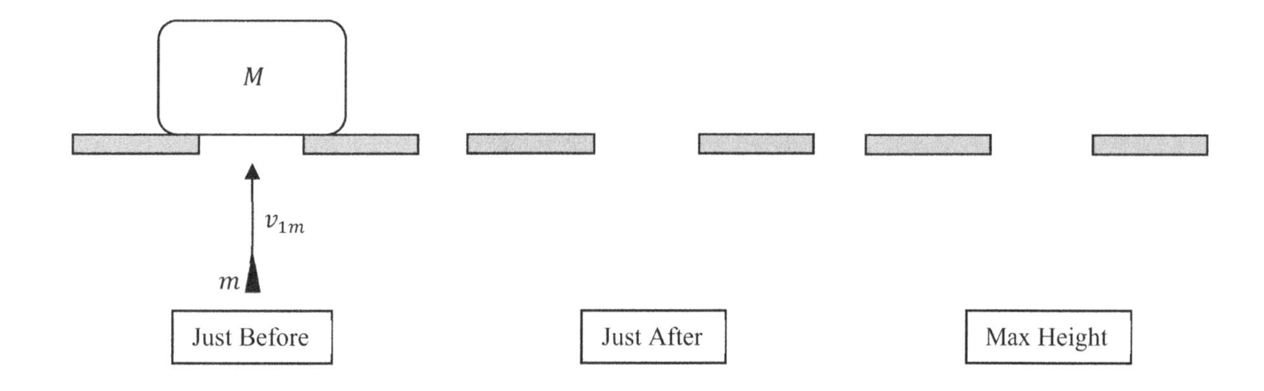

9.11 As seen from above, a truck (mass m_1) travels with initial speed v_{1i} due east. A car (mass m_2) travels with speed v_{2i} due north. After a collision the two objects stick together and slide off with unknown speed v_f along level ground for a distance d before coming to rest.

a) What type of collision is this?
b) Determine algebraic expressions for the *components* of final velocity of the combined object after the collision.
c) Determine an algebraic expression for the final speed and heading after the collision.
d) Now assume the truck has mass 800 kg and the car has mass 600 kg. Assume the truck's speed before the collision was 25 m/s while the car's speed before the collision was 40 m/s. Determine the final *velocity*. Assume masses all numbers have 2 sig figs.
e) Double check: initially, was there more momentum in the vertical or horizontal direction? Should the final velocity be closer to vertical or to the horizontal?
f) Now consider the side view showing the combined object just after the collision sliding to a halt across level ground. Determine an equation relating stopping distance d to the coefficient of kinetic friction μ.
g) As μ gets bigger, what should happen to d? Verify with your equation.

1
2
Top View, Before
v_f
ϕ
Top View, After
Comes to rest
v_f
Side View, After

9.12 Two cars are driving with constant speed towards a junction in the road as shown in the figure. Car 1 has mass m and initial speed v. Car 2 has mass $2m$ but we don't know it's initial speed. The drivers do not notice each other and as a result they collide at the location indicated by the black dot. The collision is perfectly inelastic. After the collision the cars move in unison with speed $0.441v$. Determine the initial speed of car 2. Express your answer as a number with three sig figs times v.

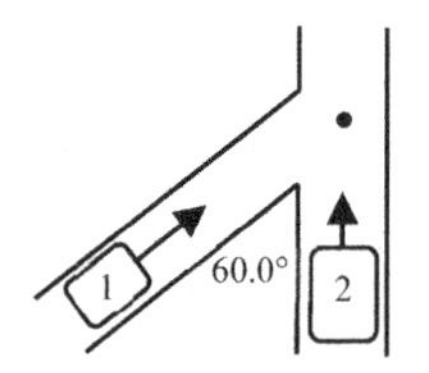

9.13 This case models an extremely over-inflated basketball dropped with tennis ball on top of it. When the overinflated basketball hits the ground it reverses direction and smacks into the tennis ball directly on top of it (before picture). After it smacks into the smaller tennis ball above it will slow down and impart some of its momentum to the tennis ball (after picture). Note: in real life there is no separation between the two balls during impact with the floor but this simplified model helps us conceptualize the problem.

a) If initially released from height h, what is the speed of the balls as the reach the ground? Ignore the size of the balls compared to the height h.
b) Why is it reasonable to assume that in the before picture the basketball ($3m$) is moving up with the same speed v that it had just before it reached the ground? What are we assuming about the collision between the basketball and the ground (elastic or inelastic)?
c) Just after the collision, what are the speeds of the two balls?
d) After the collision, what is the max height reached by the tennis ball?

9.14 Fun demo, skip problem Newton's cradle. From experience you have probably know that lifting N balls on the left causes N balls on the right to rise after the collision. The purpose of this problem is *prove* that result. Assume each ball has mass m. Assume N balls are raised and impact the rest of the chain with speed v. Assume after the collision the momentum is $N_f m v_f$ where N_f is the final number of balls raised & v_f is the speed of those balls just after the collision. Furthermore, for steel on steel, we may assume the collision is elastic.

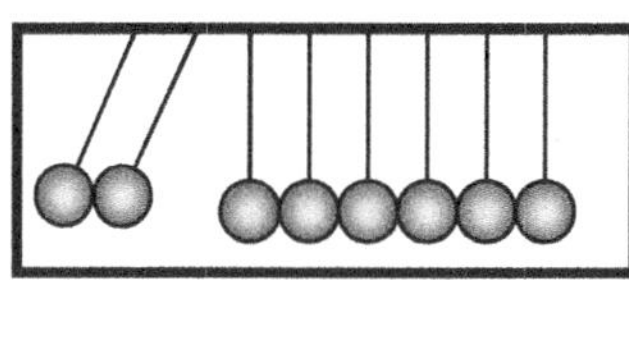

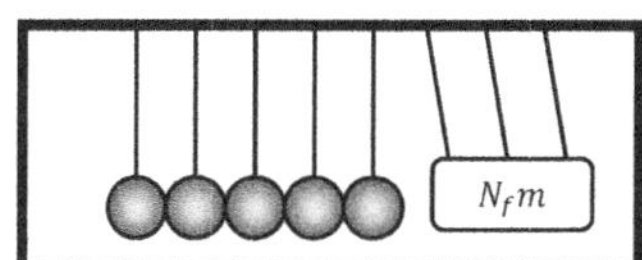

a) Determine N_f and v_f.
b) What should happen if you raise 3 balls or 4 balls?
c) What if you drop 3 from one side and 1 from the other?

Think you should try building a large Newton's cradle yourself? I have tried myself and seen numerous other attempts both online and in person. For some reason I have never seen a large version work as well as the little ones. The beautiful patterns seem to get sloppy after just a few collisions.

Note: in practice, our momentum problem only makes sense if we consider collisions <u>between 2 balls at a time</u>.

<u>If all the balls are touching</u> at once during collisions, it is possible to have all of the balls move just a little bit (rather than the same number of balls move off with identical velocity).

Note: Most textbooks work out a sweet derivation that handles head-on elastic 1D collisions of 2 objects for both objects initially moving. After doing a few problems, read this derivation and see if you can do it yourself. Knowing this derivation and its results can be a time saver during 1D elastic collisions. The derivation is similar to problem **9.7**.

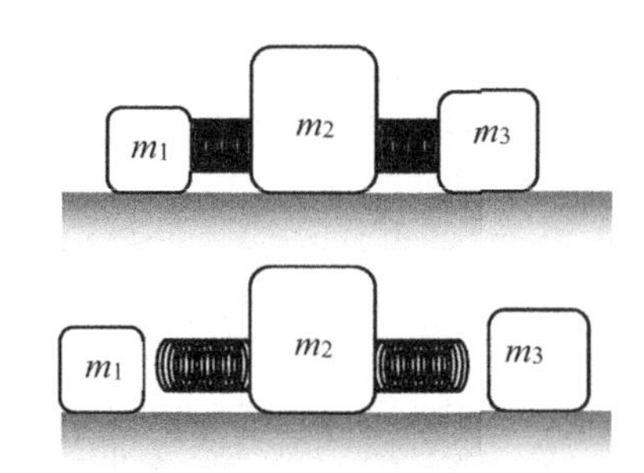

9.15 Consider the figure at right. All three blocks are initially at rest on a frictionless, horizontal surface. The blocks have masses of $m_1 = 1.00$ kg, $m_2 = 10.0$ kg, and $m_3 = 3.00$ kg. Between m_1 and m_2 a spring of constant $k = 38.125$ kN/m is compressed by 4.00 cm. An identical spring is compressed the same amount between m_2 and m_3. The system is initially held in place by hands not shown in the figure. The hands then release the blocks from rest. The blocks separate and move away from each other. Note: the springs have negligible mass. After the release m_1 moves with a speed of 8.00 m/s (relative to the earth). Notice the springs are attached to m_2 but not the other blocks after the release.

a) Is energy conserved in this experiment or not? Circle the best answer.

Yes	No	Only kinetic energy is conserved	Impossible to determine

b) Is momentum conserved in this experiment or not?

c) Which direction, if any, will m_2 travel?

d) Determine the final velocities of blocks 2 and 3. Express them as speeds and directions.

9.16 It is a cold day in the northland…In Floodwood, MN the streets are covered with a thick layer of ice. A car of mass m is trying to stop for a light. A car of mass $2m$ sits at rest at a stop light. Just before impact the lighter car is travelling with speed v. During the collision the friction from the streets acting on the cars is negligible. Furthermore, the speed of the lighter car is slow enough to allow the bumpers to do their designed job. The bumpers will act like perfect springs and allow the collision to be nearly elastic.

a) Take a guess: after the collision should each car have half the *momentum*, will each car half the *energy*, or something else? Should one of the cars stop just like when a cue ball hits another pool ball head on?

b) Determine the final speed of each car.

c) Determine the final momentum and energy of each car.

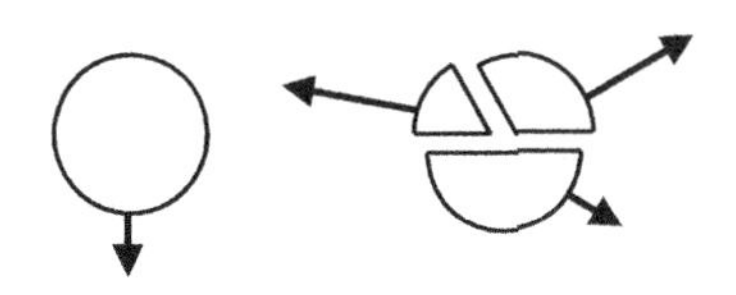

9.17 A bomb of mass $6m$ is falling to earth. At the instant just before exploding the bomb is moving downwards with speed v. Just after the explosion the bomb splits into three fragments. The fragments have mass m, $2m$, and $3m$ respectively. The fragment of mass $2m$ moves speed $2v$ directed 30° *above* the positive x-axis. The fragment of mass $3m$ moves speed v directed 30° *below* the positive x-axis.

a) What is the final speed and direction of the third fragment? Note: in the figure I took a guess that it would move to the left and *up*. Perhaps it moves to the left and *down*. You may need to think about how the signs of your answers affect the physical interpretation of your result as in problem **9.4**.

b) How much kinetic energy was added during the explosion?

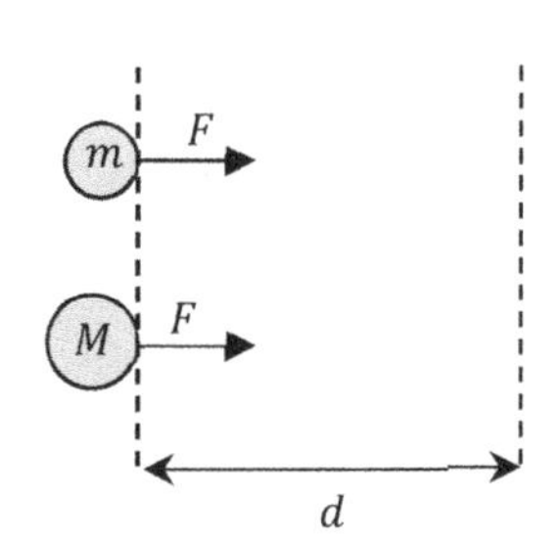

9.17½ Concept Review Question: Two masses are initially at rest at the starting line. Mass m is smaller than mass M. At time $t = 0$, each mass has a constant force of magnitude F applied for distance d. Tip: to make this problem fun, take your initial guess before thinking too much. Then, let $M = 2m$ and do the math algebraically!

a) Which mass crosses the finish line first?

b) Which mass has more *kinetic energy* at the finish line?

c) Which mass finishes the race with greater *speed*?

d) Which mass finishes the race with greater *momentum*?

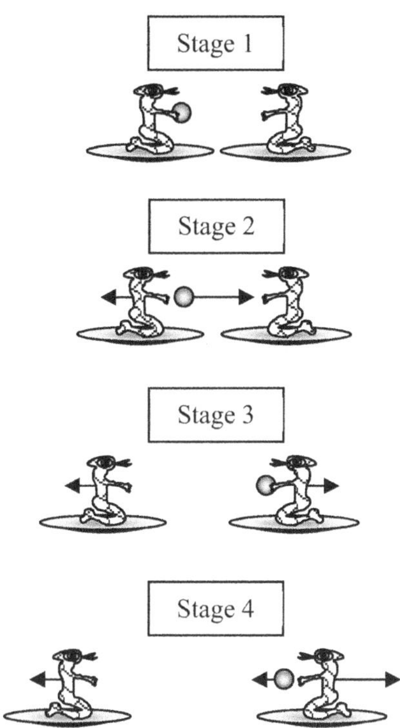

9.18 CHALLENGE Some aliens intend on throwing a medicine ball of mass m_B back and forth. The mass of each alien (including hovercraft) is m_A. Each alien will always throw the ball with speed v *relative to herself.*

In stage 1 the aliens are initially at rest.

In stage 2 the left alien throws the ball towards the right alien. Notice the left alien recoils backwards.

In stage 3 the right alien catches the ball and begins moving backwards. The left alien continues to move backwards.

In stage 4, the right alien throws the ball back towards the left alien.

Comments: none of the velocity arrows are drawn to scale. External forces are negligible during all stages. The pictures are shown from the perspective of a stationary observer watching the game of catch. If you get stuck, skip parts a) through c) and start with part d).

a) Relative to the stationary observer, how fast is the right alien moving just after *the catch* in stage 3?

b) Relative to the stationary observer, how fast is the right alien moving just after *the throw* in stage 4? Relative to the stationary observer, how fast is the ball moving just after *the throw* in stage 4?

c) Will the ball make it back to the left alien or not? Under what circumstances will the ball make it back to the first alien?

d) How long (how many tosses) can the game go on? In theory could it go on forever? Does it depend on the mass ratio $\frac{m_B}{m_A}$? I'm not going to do this but it sounds interesting.

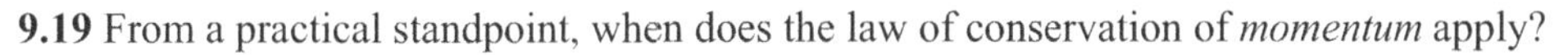

9.19 From a practical standpoint, when does the law of conservation of *momentum* apply?

9.20 In general, when does the law of conservation of *energy* apply?

9.21 Explain why the following situation is impossible. Show work to support your reasoning.
Deep in space two asteroids interact in what is essentially an elastic, head-on collision. This occurs when they come close to each other; they exert gravitational forces on each other even though they do not make contact. The first asteroid has mass 25 kg and moves with speed 200 m/s to the right. The second asteroid has mass 20 kg and is moving with speed 150 m/s to the left. After the collision the 25 kg mass moves to the left at only 140 m/s. Meanwhile, the 20 kg asteroid now moves to the right at 275 m/s.

9.22 An open box slides across the frictionless, icy surface of a frozen lake. What happens to the speed of the box as water from a rain shower falls vertically downward into the box? You may assume the icy surface remains frictionless as the box fills with water. Clearly explain why for credit.

9.23 An ice cube of mass m is placed on a ramp of mass $5m$. The block and ramp are initially at rest on a hockey rink. The block is initially distance h above the hockey rink. Upon being released from rest, the block slides down the ramp and moves off to the right. There is negligible friction at all interfaces (block and ramp, ramp and hockey rink, block and hockey rink). Because friction is negligible, the ramp recoils to the left.

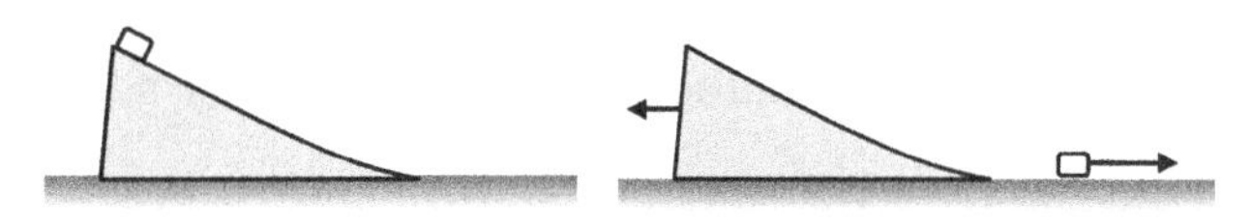

a) Determine the speed of the ramp (relative to the earth) after the block has left the ramp.

b) Determine the speed of the ice cube (relative to the earth) after it has left the ramp.

c) Determine the speed of the ice cube *relative to the ramp* after it has left the ramp.

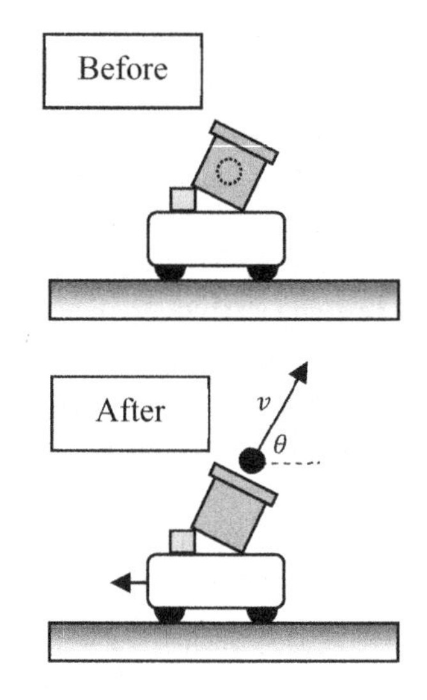

9.24 An angled spring gun is mounted on a cart on a level track. Effectively, the spring gun-cart object experiences negligible friction. The gun is triggered electronically such that no one holds the gun while it is fired. A stationary observer sees the bullet (black circle) leave the gun with speed v angled θ above the horizontal relative to the *earth.* Assume combined mass of spring gun and cart is M while mass of the bullet is m.

a) Why is it pointless to do conservation of momentum in the *vertical* direction?
b) Determine the recoil speed of the cart (relative to the earth).
c) Do you think it is reasonable to use conservation of *energy* for this problem or not? If yes, what is the energy equation? If no, why not?
d) How would things change if the wording in the problem was changed to "A stationary observer sees the bullet (black circle) leave the gun with speed v angled θ above the horizontal relative to the *cart.*"

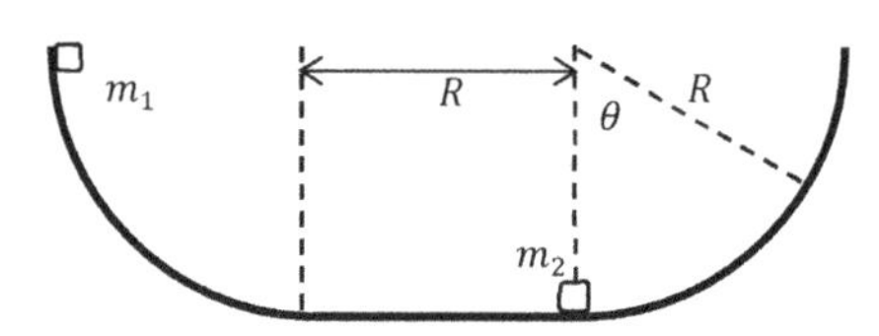

9.25 A block of mass m_1 is released from rest at the top of a quarter circle of radius R. A the bottom of the quarter-circle it slides across a level surface for distance R before impacting a second block of mass m_2. Assume friction is negligible between each block and all surfaces. Assume block size is negligible compared to the radius R. **Try at least part a & b. If desiring punishment, continue.**

a) If the collision is *perfectly inelastic*, to what max angle θ will the blocks rise on the opposite quarter circle.
b) Now assume $m_2 = 2m_1$. Also assume the collision is now *elastic*.
   i. Determine the max angle θ reached by m_2.
   ii. Determine the max angle reached by m_1 after the collision. Specifically state if this max angle is on the right or left quarter circle.
c) Once again assume the collision is *elastic*. Now assume $m_1 = 2m_2$.
   i. Determine the speed of m_2 as it launches off the right end of the track.
   ii. Determine the max height above the end of the track reached by m_2
   iii. Determine the max angle reached by m_1 after the collision. Specifically state if this max angle is on the right or left quarter circle.
d) **Challenge:** Assume kinetic friction is present on only the flat portion of the track. Assume $m_2 = 4m_1$. Assume m_2 reaches max angle of $15°$ after an elastic collision. Note: for any $m_1 < m_2$ it turns out m_1 reverses directions upon impacting m_2 during an elastic collision with m_2 initially stationary. After the impact with m_2, at what location will m_1 stop. In particular, will it stop prior to returning to the quarter circle on the left or will it reach some max angle on the left.
e) **Quickie:** Assume the two masses are identical and the track still has no friction. Describe the behavior the system after the first mass is released from rest. To guide you through this, first consider the speed of each mass after the collision. Then think about the max angle reached by m_2 after the first collision. Then think about the collision that occurs when m_2 comes back down. And the next collision...and the next...

9.26 Challenge: Write a computer simulation which correctly models the following. Imagine a flat track (like an air track) and assume a glider of mass m_1 is launched with initial speed v at a stationary glider of mass m_2. Assume the length of the track is $L = 2.00$ m. Assume collisions between any glider with either end of the air track reverses the direction of that glider without any loss of energy. Assume no friction. Assume all collisions between the two gliders are elastic. Allow users to input the glider masses between 0.200 to 0.400 g. Allow users to input v between 0 and 1.0 m/s. If you get your code written, we can compare it to reality with an air track.

Center of mass strategy:

1. Use symmetry if possible to find x_{CM} or y_{CM} without calculating anything
2. Memorize answers for common uniform shapes (usually at the center of a rod, square, circle, sphere or 1/3 from the fat end of a triangle)
3. If an object is comprised of several known shapes, split the object up and use $x_{CM} = \frac{m_1 x_1 + m_2 x_2}{m_1 + m_2}$
    a. in this equation, each x_i is the CENTER OF MASS of the ith object.
    b. One can use *negative mass* for objects with holes in them (or *negative density*)
4. If you have to use calculus, slice the object at some random spot and label it dm
    a. If slice gives you a dy somewhere it will ultimately help you find y_{CM}
    b. If slice gives you a dx somewhere it will ultimately help you find x_{CM}
    c. Use the appropriate density equation to relate dm to dx, dy, ds, dA, or dV
        i. If 2D or 3D, look at your slice to write down an equation for dA or dV
        ii. Be careful: dA or dV likely requires equation relating length, radius, etc to x or y
    d. If no comment is made about non-uniform density, assume the object is uniform. Otherwise plug in the equation for non-uniform density
    e. Use the center of mass integral $x_{CM} = \frac{1}{m}\int x\, dm$
    f. For non-uniform usually do side problem to find mass using $m = \int dm$ and check units.

# of dimensions	**name for density**	**if density <u>uniform</u>**	**types of object**	**examples of uniform densities**	**mass of slice (dm)**
1D	λ =linear mass density in some books μ =linear mass density	$\lambda = \frac{total\ mass}{total\ length}$	thin rods, wires, thin rods bent into arcs	rod: $\lambda = \frac{M}{L}$	$dm = \lambda dx$ $dm = \lambda dy$ $dm = \lambda ds$ for arcs $ds = R d\theta$
2D	σ =area mass density	$\sigma = \frac{total\ mass}{total\ area}$	Thin plates (triangular, circular, rectangular, etc)	rect: $\sigma = \frac{M}{LW}$ circ: $\sigma = \frac{M}{\pi r^2}$	$dm = \sigma dA$ slice uses dy to get y_{CM} slice uses dx to get x_{CM}
3D	ρ =volume mass density	$\rho = \frac{total\ mass}{total\ volume}$	spheres, rectangular solids, plates with varying thickness, frustums [**<u>not</u>** frustRUM] of cones	sphere: $\rho = \frac{M}{\frac{4}{3}\pi r^3}$ cylinder: $\rho = \frac{M}{\pi r^2 h}$	$dm = \rho dV$ slice uses dy to get y_{CM} slice uses dx to get x_{CM}

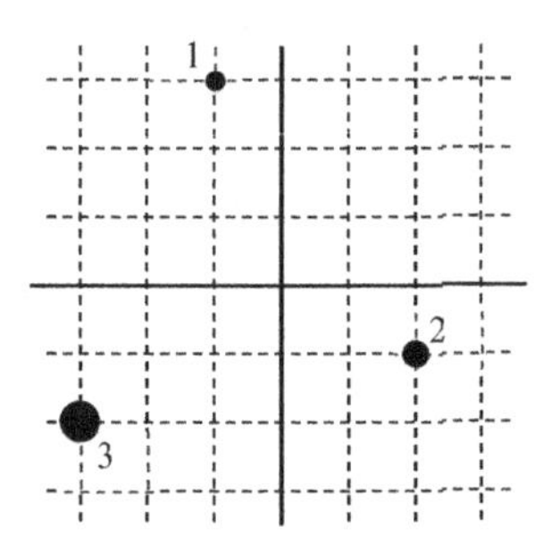

9.27 Three masses are at the positions shown. Determine both x_{CM} and y_{CM} relative to the origin shown. Assume each tick mark is 20 cm. Assume the masses are 1 kg, 2 kg, and 3 kg respectively.

a) Take a guess! Where do you think the center of mass is for this system?
b) Compute the location of the center of mass. Determine both the x- and y-coordinate. Use your result to check your intuition from the previous step.

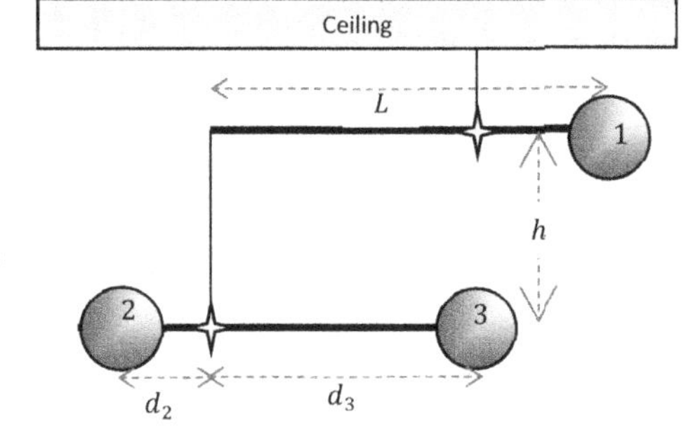

9.28 A mobile is constructed out of three bowling balls and support rods. The rods (thick black lines) have negligible mass. For the mobile to balance, the center of mass of each rod must be located at the point where a string connects to the rod. These points are shown as the four-pointed stars in the figure. Two balls have known weights ($w_1 = 16$ lbs & $w_2 = 8$ lbs). The mobile is designed such that $d_3 = 56$ cm and $d_2 = 42$ cm. The mobile is designed such that the third bowling ball will be distance h directly under the string tying the mobile to the ceiling.

a) Determine the weight required for the third bowling ball.
b) Determine the length of the upper rod L.

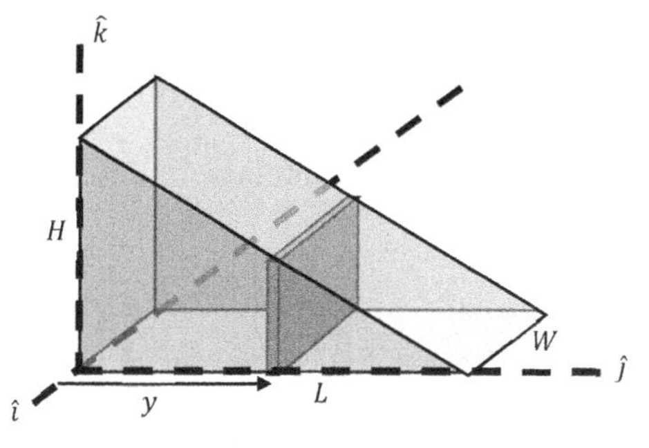

9.29 3D Calculus A uniform, *right*, triangular wedge has height H, length L, and thickness W as shown. The triangle has mass M. A representative slice of the object is shown distance y from the origin.

a) Based on the orientation of the slice shown, do you expect me to ask about x_{CM}, y_{CM}, or z_{CM}?
b) Determine the density ρ of the slice in terms of the givens.
c) Determine the height z of the slice as a function of y.
d) Determine the mass of the slice dm in terms of only M, L and y.
e) Determine the horizontal location of the center of mass.
f) Think: how much mass is to the left and right of the center of mass? Is more mass on one side or the other of the center of mass or should the mass be the same on either side of the center of mass?
g) What is the (x, y, z) coordinate of the center of mass?

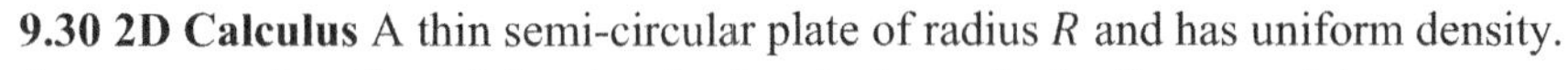

9.30 2D Calculus A thin semi-circular plate of radius R and has uniform density. A representative slice of the plate is already drawn in the figure at right.

a) Based on the slice shown, do you expect me to ask about x_{CM} or y_{CM}?
b) Should we use λ, σ, or ρ for density in this case?
c) In terms of only y and R, determine the area of the slice.
d) Determine the mass of the slice in terms of density, R, and y.
e) Determine the position of the center of mass.
f) Determine the total mass of the plate in terms of density and R.
g) **Challenge:** How much mass is above and below the center of mass? Is more mass above, below, or is it a tie?

Potentially useful integral for part e

$$\int \sqrt{a^2 - x^2}dx = \frac{1}{2}x\sqrt{a^2 - x^2} + \frac{1}{2}a^2 \tan^{-1}\frac{x}{\sqrt{a^2 - x^2}}$$

9.31 Assume the shape has uniform density. You may assume it is a 2D plate. You are asked to use the coordinate system shown by the thick dashed lines.

a) Show the x-coordinate of the center of mass is given by:

$$x_{\text{CM}} = \frac{\frac{ad^2}{3} - (a+b)c^2}{2(a+b)c + ad}$$

Notice this is ugly no matter how you try to simplify it.

b) Assuming the figure is drawn to scale, do you expect x_{CM} to be a positive or negative number?

c) Suppose you are now told an architect wants the same type of shape but she wants x_{CM} to be above the origin (for cantilevering purposes). She says you are free to adjust parameter only the parameter d. To make x_{CM} line up above the origin, how will d compare to c? Without doing any math, first take a guess: $d = c$, $c < d < 2c$, $d = 2c$, $d > 2c$? Use math to show $d = c\sqrt{3\frac{a+b}{a}}$. Then consider the special cases of $b = 0$ and $b = a$. Was your intuition correct for these special cases? I know it surprised me when I first did it!

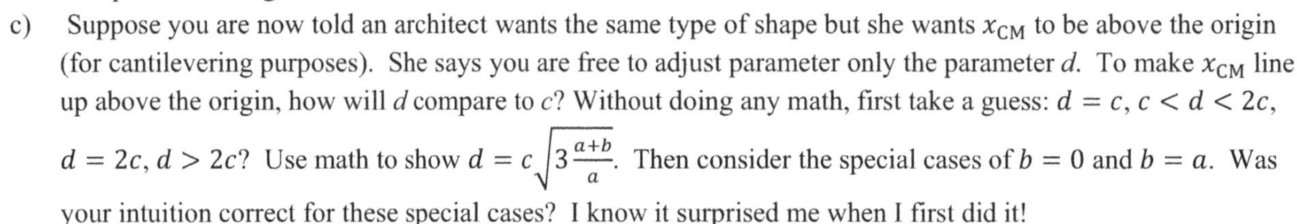

9.32 Consider the thin circular plate with a square hole cut out of it (see figure). The radius of the circle is R. Determine the location of the x_{CM}. I found it surprising how little that large hole shifted the center of mass…

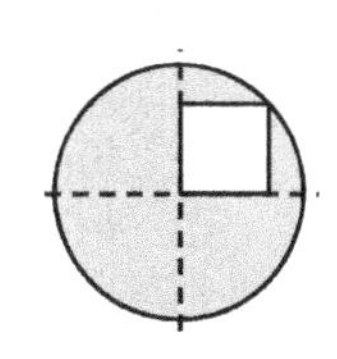

9.33 1D Calculus Suppose a rod with length L has linearly increasing density. **This is an example of non-uniform density.** Based on the coordinate system shown, density is given by $\lambda = \alpha x$.

a) Based on the density function, where do you expect the center of mass to be: in the middle, towards the right end, or towards the left end? Hint: which end of the rod should be heavier based on λ?
b) What are the units of α?
c) Determine the total mass of the object in terms of α and L. Total mass is given by $M = \int dm$. Remember to check the units. The answer to part b) is used to help you check your units to your answer for part c).
d) Determine the horizontal location of center of mass of the object in terms of L (is it $L/2$, $L/3$, etc).
e) Think: compare your result to a right triangle with the point at the origin. *Why* are the results the same?

9.33$\frac{1}{3}$ Suppose the previous questions was modified. You may still assume the total length of the rod is L. This time, however, the rod is centered on the origin. Furthermore, the density is changed to $\lambda = \alpha x^2$. We can tell $x_{\text{CM}} = 0$ by symmetry, but still consider the following.

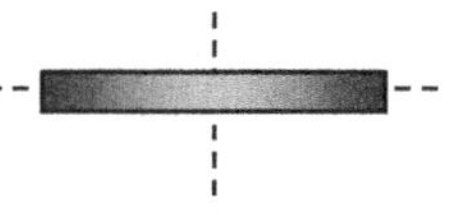

a) What would change from the previous work? Density equation changes, but would anything else change?
b) Why is it physically unreasonable to use $\lambda = \alpha x$ as the density equation for this scenario?

9.33$\frac{2}{3}$ A non-uniform, thin rod has a mass distribution that is given by $\mu(x) = ax^n$. Here n is unknown number (greater than 1) and a is an unknown constant. This density assumes the coordinate system is aligned with the left end of the rod as shown in the figure. The rod has total length $L = 1.00$ m and mass $m = 0.100$ kg. By balancing the object, you determine the center of mass is 14.29 cm from the right end of the rod (shown by the black **x** in the figure). Figure not to scale. Determine the power of n. Express your answer as a decimal with 3 sig figs.

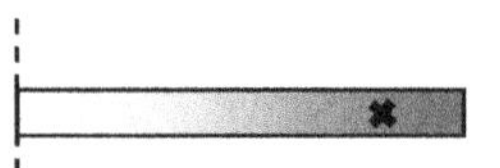

9.34 Non-uni class demo One way to create a rod with non-uniform density is to take a piece of wood and attach strips of aluminum to the top. Each strip is a slightly different length. In the end the rod will have an effective density that increases from left to right. I made such a rod. I determined the density at the *left* end of the rod is λ_L. The density at the *right* end of the rod is λ_R. Rod length is L.

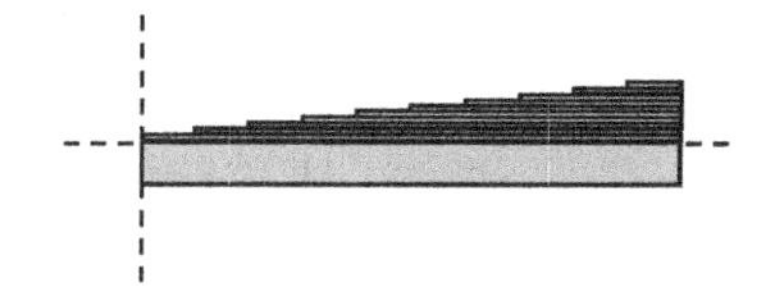

a) Determine density of the rod as a function of x assuming it varies *linearly*.
b) Determine the center of mass in terms of λ_L, λ_R, and L.
c) The demo has a length of 50.0 cm, $\lambda_L = 3.62\ \frac{\text{g}}{\text{cm}}$, and $\lambda_R = 13.14\ \frac{\text{g}}{\text{cm}}$.
   For these values I found $x_{\text{CM}} = 29.7$ cm from the left end.
   Think: should the number be greater than or less than 25.0 cm?
   Does my answer make sense?

9.35 2D Calculus Consider a thin, triangular plate with height H and length L. The non-uniform density is given by $\sigma = \alpha - \beta x^2$ based on the coordinates shown. This problem is obviously tedious, but each step is fairly straightforward. Rather than finishing this problem, show you get to a step where

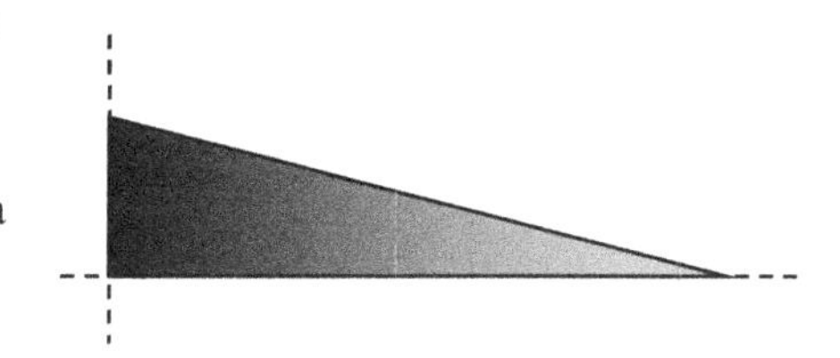

$$x_{\text{CM}} = \frac{\int_0^L (\alpha - \beta x^2)(Lx - x^2)dx}{\int_0^L (\alpha - \beta x^2)(L - x)dx}$$

9.36 Consider the uniform disc with attached rods shown at right. We are able to vary the angle between the rods. The disc has mass m_1 and radius R while the rods each have mass m_2 and length L. What angle θ allows the system to be balanced on the edge of the disc? With a little thinking, you might be able to convince yourself this occurs at the point shown by the black × in the figure.

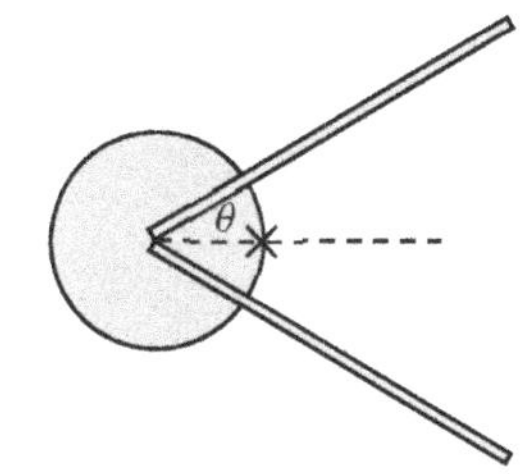

9.37 2D Calculus Consider the shape shown in the figure at right. The top curve is known to be a parabola. The parabola's minimum occurs at the left end. The plate has uniform mass density of σ.

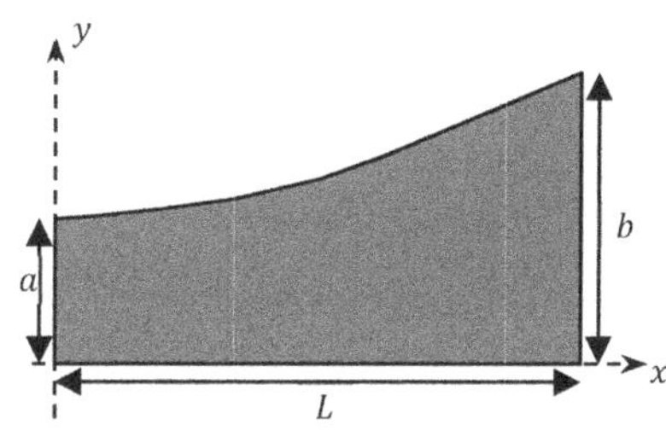

a) Determine an expression for $y(x)$ in terms of a, b, & L. Use the general equation for the parabola $y(x) = c_0 + c_1 x + c_2 x^2$. The left boundary point gets you one of the constants. The right boundary gets you an equation relating the other two constants. We also know the *minima* occurs at position $x = 0$…the derivative of $y(x)$ gets this condition involved.
b) Show the plate has mass $m = \frac{\sigma L a}{3}\left(2 + \frac{b}{a}\right)$. Think: does this result make sense when $b = a$?
c) Show the x-coordinate of the center of mass is $x_{\text{CM}} = \frac{3L}{4}\frac{(a+b)}{(2a+b)}$. Think: does this make sense when $b = a$?
d) **Challenge:** Show $y_{\text{CM}} = b\frac{\left(3+4\frac{a}{b}+8\left(\frac{a}{b}\right)^2\right)}{\left(50-20\frac{a}{b}\right)}$. Think: does the result make sense when $b = a$? Hint: consider splitting the shape up as a rectangular plate with a parabolic hole as shown at left. This makes the integrals easier.

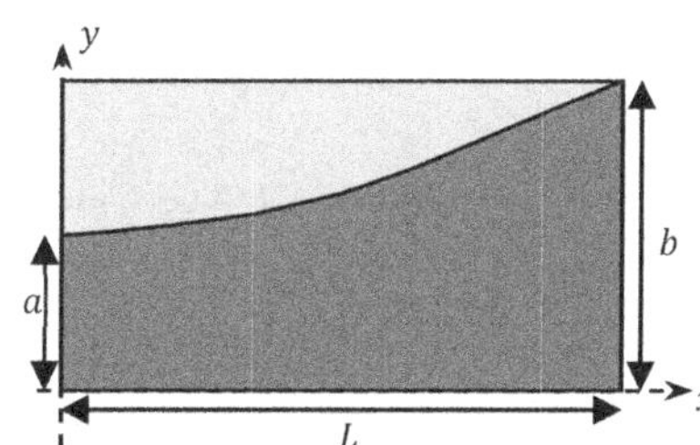

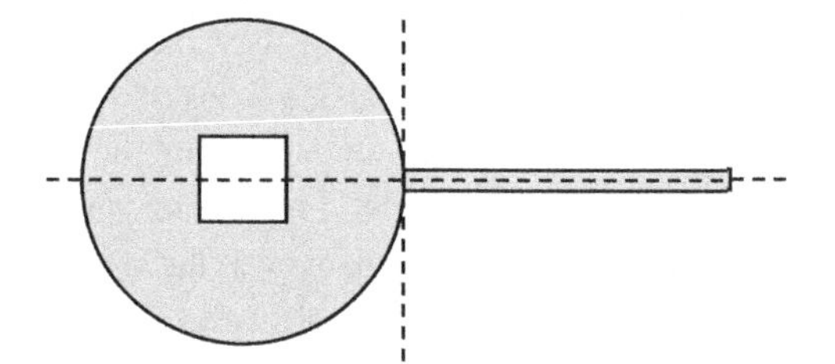

9.38 1D Calculus A thin rod is connected a circular metal plate with a square hole in it. The circular plate has uniform mass density σ and radius R. The square hole is centered on the center of the circular plate and has a side of length $s = R/4$. The rod has length $L = 2R$ and has a mass distribution given by $\lambda = ax^{1/3}$.

a) Determine the units of a.
b) Determine the total mass of the non-uniform thin rod in terms of a and R.
c) Determine the center of mass $(x_{\text{rod}}, y_{\text{rod}})$ of the non-uniform thin rod.
d) Determine an expression for the mass of the circular plate (without the hole) in terms of R and σ.
e) Determine an expression for the mass of the hole in terms of R and σ. Note: since it is a hole it has negative mass. This negative mass will subtract away some of the plate.
f) Combine all three pieces to determine $x_{\text{CM}} = \frac{x_1 m_1 + x_2 m_2 + x_3 m_3}{m_1 + m_2 + m_3}$. Final answer will include a, σ, and R.
g) Think: if the hole was made larger would the center of mass shift to the right or to the left?
h) CHALLENGE: If the mass distribution of the rod was instead $\lambda = bx^{1/2}$, which way would x_{CM} shift?

9.39 3D Calculus/Challenge: determine the z_{CM} coordinate of a hemisphere?
What would you do if it was non-uniform?

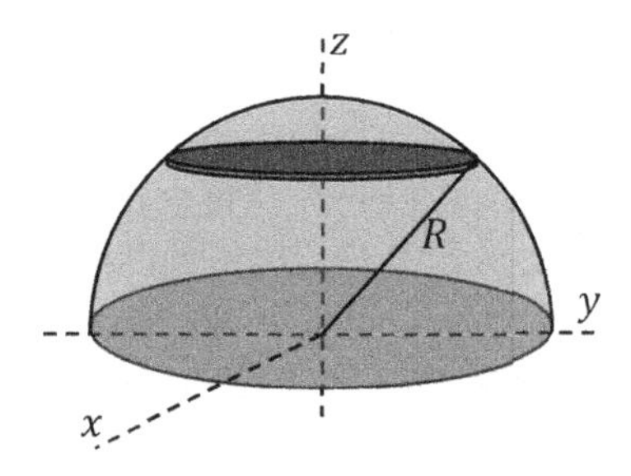

9.40 3D Calculus/Challenge: Consider the frustum of a cone (truncated cone) shown at right. Set up the integral to find the center of mass.
I found

$$x_{\text{CM}} = H\frac{R_1^2 + 3R_2^2 - 2R_2R_1}{4(R_1^2 + R_2^2 + 2R_2R_1)}$$

I wasn't sure if this answer was correct.

One way to check this result is to compare it to the known formula for a right cone.
For a right cone, the center of mass is along the axis ¼ of the height from the base.
Verify the above equation gives that result.
Hint: by separately considering $R_1 = 0$ versus $R_2 = 0$ you can check your formula in two distinct ways.

Can you think of an even easier way to check this formula?

Note: the answer above *may* or *may not* be correct…

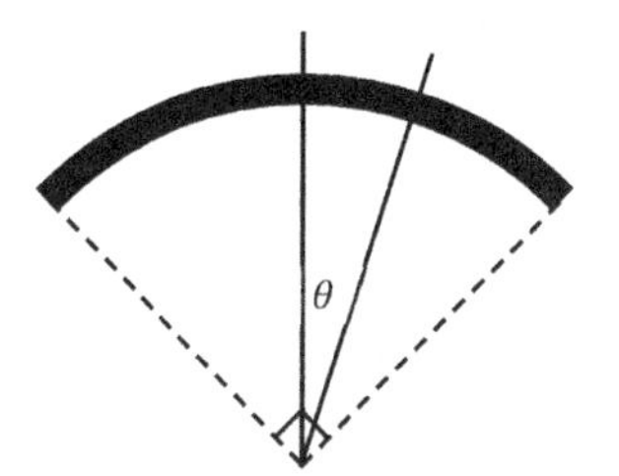

9.41 1D Calculus Consider a thin rod of mass m bent into a quarter-circle of radius R as shown in the figure. This is a crude model of a high jumper doing the Fosbury flop. For this orientation, I typically assume purely vertical is $\theta = 0$. Angles *clockwise* from vertical are *positive* while angles *counterclockwise* from vertical are *negative*. Also, because we intend to do calculus, we must express angles in radians. The angles in this problem therefore range from $-\frac{\pi}{4}$ to $+\frac{\pi}{4}$.

- For thin rods bent into an arc $dm = \lambda ds = \lambda R d\theta$.
- For *this* choice of coordinates $y = R\cos\theta$ and $x = R\sin\theta$.
- You should find $y_{CM} = \frac{2\sqrt{2}}{\pi}R = 0.900R$.
- For a given initial jump speed (ignoring air resistance), your center of mass will rise to the same height regardless of your shape. By bending your body, some of it is lowered and the rest is raised up. Your center of mass height is not changed. When you arch your back, you are getting one body part over the bar at a time…

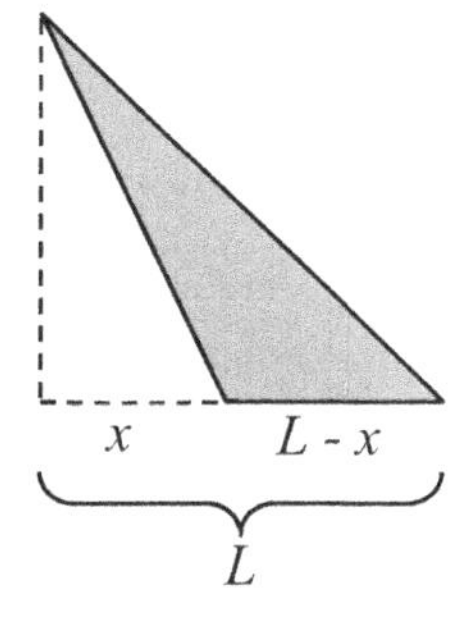

9.42 What if we have an oblique triangle? Assume the dotted lines indicate the x- and y-axes.

a) Determine x_{CM}. Hint: rather than doing a bunch of calculus, consider this as a right triangle with a right triangular hole cut out of it.

b) Think: what *should* the result be when $x = 0$. Does it make sense?

c) **Challenge:** consider the other extreme limit of $x \approx L$. What *should* this limit give as a result? Why is this basically the same thing as problem **9.33** *instead of a uniform rod*?

9.43 Zombie in the canoe:

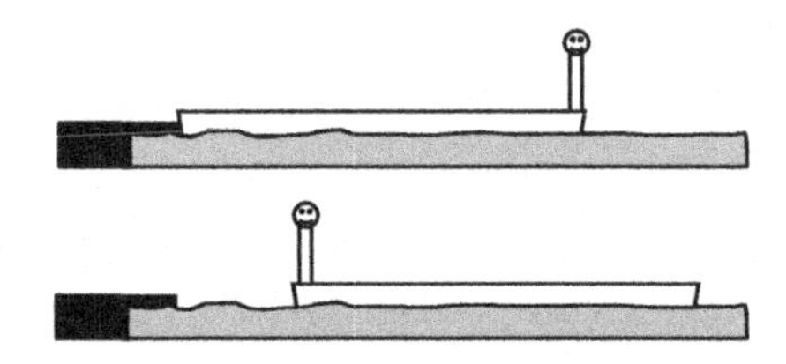

A zombie of mass m_1 stands at the right end in a canoe of mass m_2 & length L. The left end of the canoe is just barely touching the dock. The zombie walks to the left end of the canoe then stops. Drag between the canoe and the water is negligible while the man walks towards the dock.
Is the net external force zero (or negligible)? YES!
Is momentum of the zombie-canoe system conserved as the zombie walks towards the dock? YES!
If the initial momentum is zero, and momentum of the zombie-canoe system is conserved, **we know the center of mass of the zombie-canoe system is at rest!**
Said another way, if momentum is conserved

$$x_{CMi} = x_{CMf}$$

Use this information to show the zombie ends up distance $x = \frac{m_1}{m_1+m_2}L$ from the dock.
Suggested method to survive all zombie attacks? Put all zombies in canoes. Ensure zombies initially at rest at far end of canoes. Let me know how it works out.

Note: For groups of objects we can determine center-of-mass velocity and acceleration using

$\vec{v}_{CM} = \dfrac{m_1\vec{v}_1 + m_2\vec{v}_2}{m_1 + m_2}$	$\vec{a}_{CM} = \dfrac{m_1\vec{a}_1 + m_2\vec{a}_2}{m_1 + m_2}$

Why care about this last piece? During collisions $\vec{v}_{CM}$ will remain unchanged. Many physicists do collision problems in the "center-of-mass frame". Common practice in particle physics. I'll try to throw in a problem later…

9.44 A person of mass M=70 kg stands on top of a board with mass m=10 kg. Initially the person is at rest in the center of the board of length L=4.0m. The board rests on small rollers which effectively allow the board to slide without any friction over the ground. The person begins to walk to the right at the brisk pace of 2.0 m/s (a little over 4 mph). Notice immediately that this situation can only exist for a very short time period. **In the second picture, assume the person has walked halfway to the end of the board.** Figure not to scale.

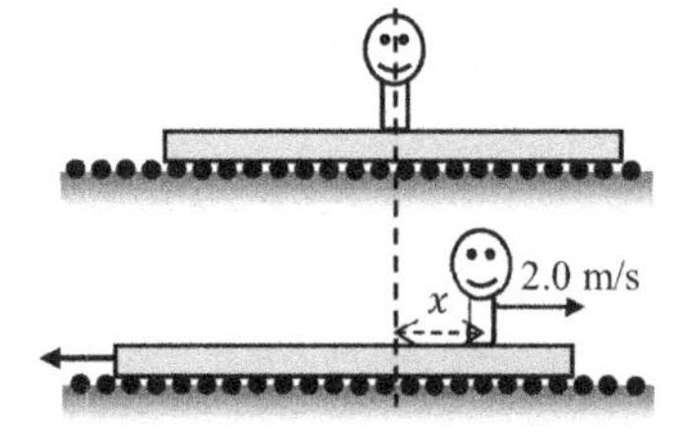

a) Is there a net external force on the board-person system?
b) Assume the center of mass in each picture is indicated by the dotted line. Based on your answer to the previous question, would you expect the center of mass to move to the right, to the left, or remain unchanged in the right picture (relative to earth which is assumed stationary)?
c) Can conservation of momentum be used for the board-person system? Explain why or why not. Hint: again consider the first question.
d) Write down an <u>algebraic</u> expression for x, the distance of the person from the center of mass. Then use numbers to compute a numerical value for x. Hint: set $x_{CM} = 0$ in the first picture.

9.45 Challenge Now comes an even trickier part. Consider the person's speed of 2.0 m/s. This is the person's speed *relative to the board* (not the earth). The person has been walking for 0.5 sec to traverse 1.0 m of plank but has only moved 0.125 m relative to the earth's surface!

a) Write down the relative velocity equation that relates the velocity of the person to the board and the earth.
b) Write down the conservation of momentum equation for the person-board system. Think: do you need to use velocities relative to the board, to the person, or to the earth? Explain.
c) Use the above two results to find algebraic equations for the velocity of the board and the person relative to the earth. Show you obtain the expected result for the person of $\vec{v}_{PE} = +0.25\frac{\text{m}}{\text{s}}\hat{\imath}$.
d) Finally, write down the algebraic expression for the velocity of the center of mass in terms of the velocities of the person and the board relative to the earth. Using your results of the previous step, show that the center of mass remains stationary (relative to the earth).

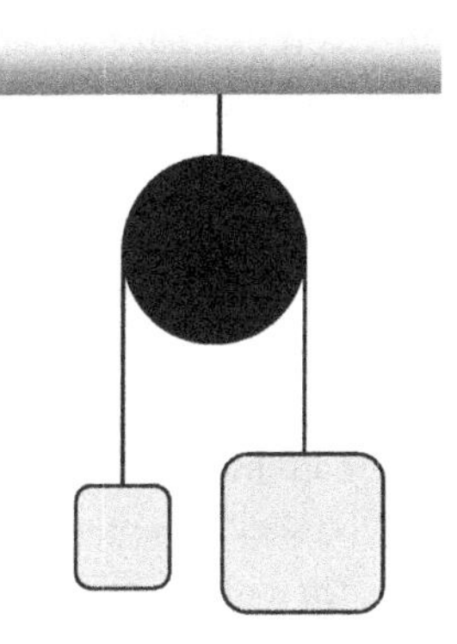

9.46 For this Atwood's machine assume the larger block has $m_2 = 2m$ while the smaller block has $m_1 = m$. The system is released from rest when both blocks are at the same initial height. Find *acceleration* of the center of mass & tension in the upper cable. The purpose of this problem is to show you the tension in the upper cable is *not* $m_{tot}g$. Using $T_{upper} = m_{tot}g$ is *approximately* true only when center of mass acceleration is negligible. Think: the masses are not being fully supported if the center of mass is accelerating downwards!

Hint: You know $y_{CM} = \frac{y_1 m_1 + y_2 m_2}{m_{tot}}$ and masses are constant.

Taking two derivatives of the center of mass equation gives

$$a_{CM} = \frac{a_1 m_1 + a_2 m_2}{m_1 + m_2}$$

Because y_{cm} is the center of mass *position*, y_{CM} is a *vector* (implied $\hat{j}$ on the result). This implies a_{CM} is a *vector* (implied $\hat{j}$ on the result). This implies you must account for the direction of a_1 & a_2 in the a_{CM} equation.

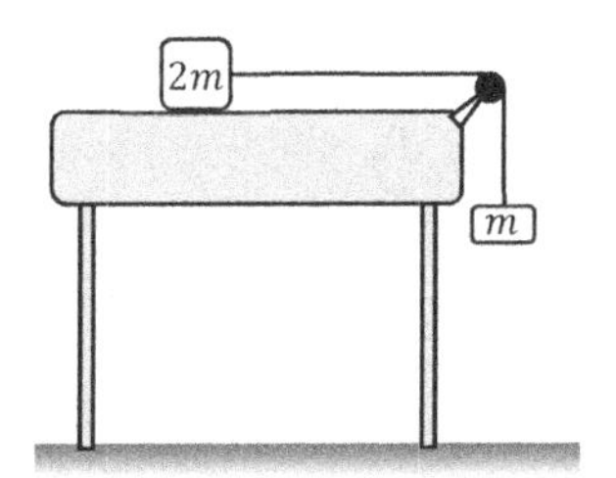

9.47 What about the Fletcher's trolley released from rest? You may assume friction is negligible. Determine x and y components of center of mass velocity and acceleration as functions of time. Hint: think about the hint in the last problem...

9.48 Excellent final exam review question A man named Moon keeps getting email intended for a man named Mooon, with *three* o's. As a token of his appreciation, Moon sends Mooon the following physics question. Moon records the motion of a 500-gram ball on his phone and uses data tracking software to record a plot of position versus time for the ball during a collision. Moon makes the position versus time plot shown at right.

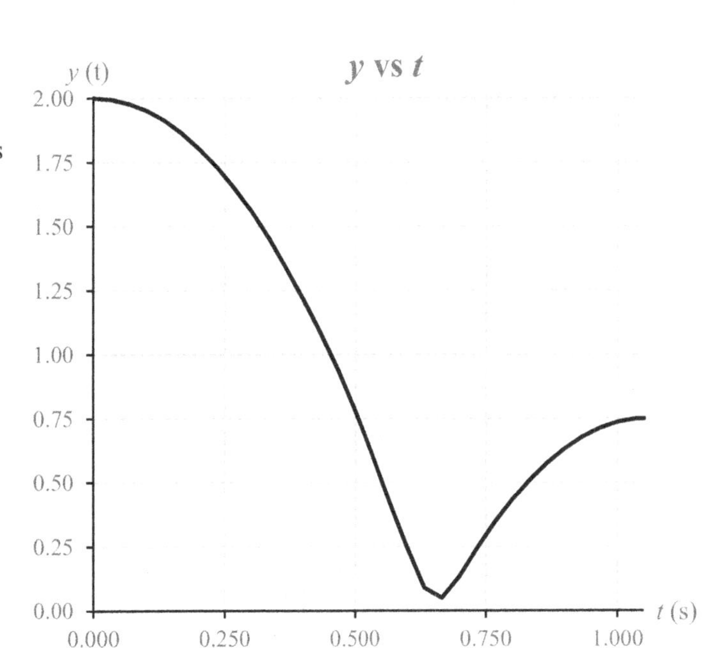

a) Estimate the speed of the ball just before impact.
b) Estimate the speed of the ball just after impact.
c) Estimate the collision time with the ground. Yours may not match mine but hopefully the thought process matches.
d) Estimate the impulse on the ball during impact.
e) Estimate the average force on the ball during impact. Yours may not match mine but hopefully the thought process matches.
f) Estimate the average acceleration of the ball during impact. Yours may not match mine but hopefully the thought process matches.
g) Sketch the plot of velocity versus time. Yours may not match mine but hopefully the thought process matches.

9.49 Rocket Propulsion

For almost every problem in this book objects are typically assumed to have constant mass. Now consider what happens if you have a rocket that uses up fuel to produce thrust. The mass of the system changes over time. To simplify the problem, let us start with a rocket in deep space experiencing negligible gravitational forces.

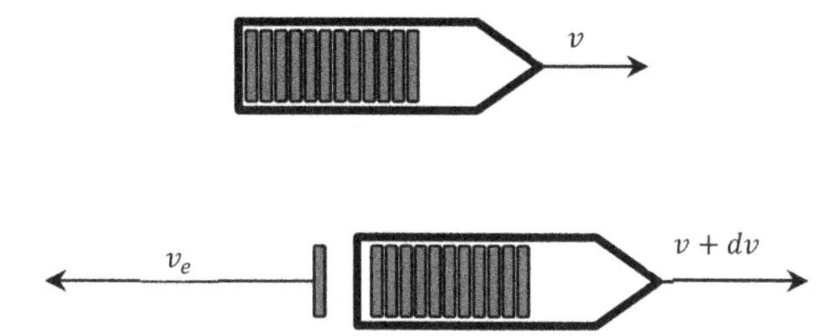

Assume the mass of the rocket and all onboard fuel is M. Notice this mass will change over time. Assume the rocket periodically ejects chunks of fuel of mass dm with constant speed v_e _relative to the rocket_. Notice when a chunk of mass dm is ejected, the change in the mass _of the rocket_ is $dM = -dm$. At one instant in time the rocket is moving with speed v. It will emit a chunk of mass dm and speed up to a new speed $v + dv$.

a) Consider an observer at rest watching the rocket go by. Do they see the fuel move backwards with speed v_e or something else?
b) Write down the conservation of momentum equation. Think carefully about the term for the chunk of fuel moving backwards. The problem statement says it is ejected with speed v_e _relative to the rocket_.
c) Eliminate the variable dm using $dM = -dm$.
d) Show the conservation of momentum equation reduces to $Mdv = -v_e dM$.
e) **Use separation of variables** to determine v as a function of M.
f) How much thrust force is generated?

9.50 Challenge Requires Calculus A chain of mass m and length L is initially held at rest such that the bottom link of the chain is just barely in contact with the floor. Determine an expression for the normal force exerted by the floor on the chain after length x has come to rest on the floor. Use your formula to determine the normal force exerted by the floor at the instant the last link hits the floor. You could test this experimentally by dropping a chain on a scale. Recall that a scale displays normal force (not weight).

9.51 1D Calculus A plastic box of mass m is pushed to the right and released. The box then slides across a frozen lake experiencing negligible friction. The initial speed of the box is v_0. Assume the sky is dumping down sleet or icy rain vertically. As a result, mass is added to the box at a rate of dm/dt. The box continues to slide with negligible friction. Determine the velocity and position of the box as functions of time.

9.52 When an object hits the earth it appears that *momentum* is not conserved. For example, a rock is dropped from a height of 10.0 m. Just before impact it is moving and has negative momentum. Just after impact it smashes into the earth and stops. Is momentum conserved or not? Explain.

Intro to Using Center of Mass Frame?

Consider two particles **1** & **2** about to collide as shown below. The figure on the left shows both particles moving relative to the stationary earth. The figure on the right views the motion from the perspective of an observer at rest relative to the center of mass (indicated by the **x**). Notice in the left picture the center of mass is moving (relative to the earth) towards the eventual collision site.

1 and **2** moving relative to earth

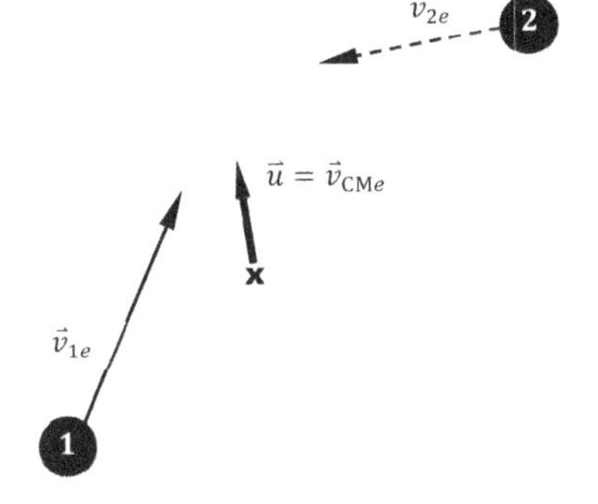

Motion relative to center of mass

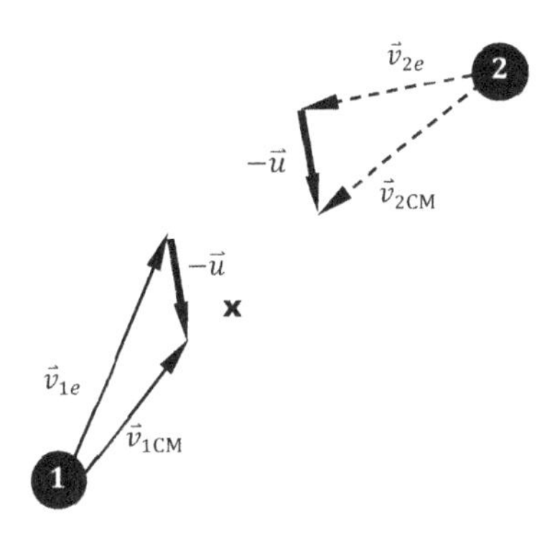

For collision problem in center of mass frame

- In before picture masses always move towards each other (towards center of mass).
- In after picture masses always move away from each other (away from center of mass).
- Center of mass remains motionless (that is essentially the definition of "center-of-mass" frame).
- Usually a scientist in a lab (an observer) is <u>not</u> motionless relative to the center of mass. To convert from the center of mass frame to the observer's frame, use relative velocity.

Using standard notion, most resources will say the initial velocities relative to the earth are $\vec{v}_1$ and $\vec{v}_2$ instead of $\vec{v}_{1e}$ and $\vec{v}_{2e}$. Similarly most resources will use $\vec{v}_1{}'$ and $\vec{v}_2{}'$ instead of $\vec{v}_{1CM}$ and $\vec{v}_{2CM}$. To reduce clutter further, the center of mass velocity (relative to earth) is shortened to $\vec{v}_{CMe} = \vec{u}$. If we consider the relative velocity of 1 we find

$$\vec{v}_{1CM} = \vec{v}_{1e} + \vec{v}_{eCM}$$
$$\vec{v}_{1CM} = \vec{v}_{1e} - \vec{v}_{CMe}$$
$$\boldsymbol{\vec{v}'_1 = \vec{v}_1 - \vec{u}}$$
$$\boldsymbol{\vec{v}_1 = \vec{v}_1{}' + \vec{u}}$$

Similarly, for particle 2,

$$\boldsymbol{\vec{v}'_2 = \vec{v}_2 - \vec{u}}$$
$$\boldsymbol{\vec{v}_2 = \vec{v}_2{}' + \vec{u}}$$

Why all this busywork? Now it is easier to write down conservation of momentum.

$$m_1\vec{v}_{1i} + m_2\vec{v}_{2i} = m_1\vec{v}_{1f} + m_2\vec{v}_{2f}$$
$$m_1(\vec{v}_{1i}{}' + \vec{u}) + m_2(\vec{v}_{2i}{}' + \vec{u}) = m_1\left(\vec{v}_{1f}{}' + \vec{u}\right) + m_2\left(\vec{v}_{2f}{}' + \vec{u}\right)$$
$$m_1\vec{v}_{1i}{}' + m_2\vec{v}_{2i}{}' = m_1\vec{v}_{1f}{}' + m_2\vec{v}_{2f}{}'$$

Notice the center of mass velocity is unchanged during the collision…that is why we don't need subscripts for $\vec{u}$! Also notice all the $\vec{u}$ terms will drop out. This means conservation of momentum is equally valid in earth's reference frame or the center of mass reference frame. Whichever frame makes it easier to compute can be used. Then, if the velocity in the other frame is desired, simply use the bold equations above. Remember

$$\vec{u} = \vec{v}_{CMe} = \frac{m_1\vec{v}_1 + m_2\vec{v}_2}{m_1 + m_2}$$

Lastly, remember that all this is only valid as long as EXTERNAL FORCES ARE NEGLIGIBLE.

Rotation

The following information applies to rotating objects (or points on rotating objects) as long as the radius is fixed.

Definitions	Useful Nuggets
$\vec{\theta}$ = angular *position* **<u>in radians (***See Note Below)</u>**	$1 \text{ rev} = 2\pi \text{ rad} = 360°$
$\Delta\vec{\theta}$ = angular *displacement* **<u>in radians</u>**	<u>Unless otherwise specified</u> $\vec{\theta}_i = 0$ and $\Delta\vec{\theta} = \vec{\theta}$
$\vec{\omega} = \frac{d\vec{\theta}}{dt}$ = angular *velocity* in rad/s	$\omega = 2\pi f = \frac{2\pi}{\mathbb{T}}$
$\vec{\alpha} = \frac{d\vec{\omega}}{dt}$ =angular *acceleration* in rad/s^2	$1\frac{\text{rad}}{\text{sec}}\left(\times\frac{60 \text{ s}}{1 \text{ min}}\times\frac{1 \text{ rev}}{2\pi \text{ rad}}\right) = 9.549 \text{ RPM} \approx 10 \text{ RPM}$
$\mathbb{T}$ = period in sec (time for 1 oscillation or revolution)	
f = frequency in Hz	$\text{Hertz} = \text{Hz} = \text{s}^{-1} = \frac{1}{\text{s}} = \frac{\text{rev}}{\text{s}} = \frac{\text{cycles}}{\text{s}}$

Angular Kinematics

The equations for angular kinematics are nearly identical to regular kinematics.

$$\Delta\theta = \omega_i t + \frac{1}{2}\alpha t^2 \qquad \omega_f^2 = \omega_i^2 + 2\alpha(\Delta\theta) \qquad \omega_f = \omega_i + \alpha t$$

Strictly speaking $\vec{\theta}$ is NOT a *true* vector. We may treat θ as one as long as the object's axis of rotation remains fixed. That said: quantities in the above equations are not magnitudes but actually scalar components which *can take on ± values*. Pay attention to the distinction between angular *speed* & angular *velocity* (or angular acceleration versus *magnitude* of angular acceleration).

Vector Nature of Rotational Quantities

For standard rotation in the xy-plane physicists use $+\hat{k}$ (out of the page) for counter-clockwise and $-\hat{k}$ (into the page) for clockwise. This may seem a bit odd but this convention will later prove useful. The convention is designed to match the right-hand rules of vector cross-products. To understand, do the following procedure:

1) Line up the fingers of your right hand with the initial angular position indicated by the dotted line.
2) Curl the fingers of your right hand to the final angular position indicated by the solid black line.
3) Stick out your right thumb and notice which way it points (right, left, up, down, into or out of the page). Your thumb lines up with the disk's *axis*.

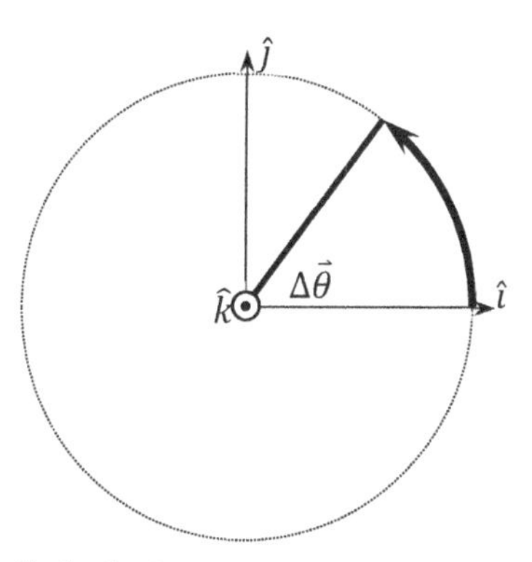

Note: *If everything in a problem is rotating clockwise, I usually flip the coordinates and call clockwise positive.*

Relating rotation to translation:

Definitions	Useful Nuggets
	$s = r\Delta\theta$
s = arclength in m (distance traveled by point along part of circular path)	$v = r\omega$
$v_{tan} = v$ = tangential speed in m/s	$a_{tan} = r\alpha$
a_{tan} = scalar component of *tangential* acceleration in m/s^2	$a_c = \frac{v^2}{r} = r\omega^2$
a_c = magnitude of *centripetal* acceleration in m/s^2	$a = a_{tot} = \sqrt{a_c^2 + a_{tan}^2}$
$a = a_{tot}$ = magnitude of *total* acceleration in m/s^2	$= r\sqrt{\omega^4 + \alpha^2}$
	These equations require angles in <u>radians</u>

If you multiply the rotational kinematics equations by r you get back translation (using s & a_{tan} instead of x and a).

$$s = vt + \frac{1}{2}a_{tan}t^2 \qquad v_f^2 = v_i^2 + 2a_{tan}s \qquad v_f = v_i + a_{tan}t$$

10.1 A record player is initially rotating at 33 1/3 RPM. The record player has radius of about 24 cm. It comes to rest in 10.0 seconds.

a) Determine the magnitude of angular acceleration.
b) Determine the magnitude of total acceleration for a point on the rim 2.00 seconds before it comes to rest.
c) Determine the total distance traveled by a point on the rim as the player slows to a stop.

10.2 A train leaves Kaohsiung Central Park headed to Sanduo Shopping District. Starting from rest, the train speeds up at rate $1.40\,\frac{\text{m}}{\text{s}^2}$ until it reaches speed $25.7\,\frac{\text{m}}{\text{s}}$. It then continues at constant speed towards Sanduo. The radius of the quarter-circle bend is $R = 20\underline{0}$ m.

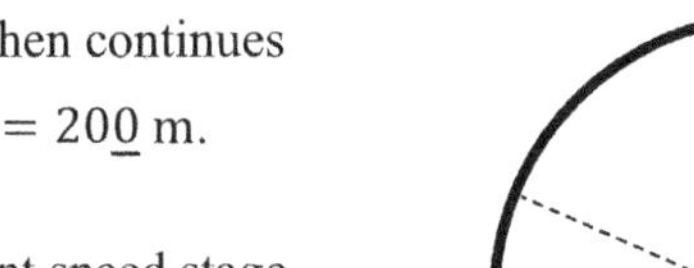

a) Determine the angle through which the train accelerates.
b) Determine the elapsed times of the acceleration stage and the constant speed stage.
c) Sketch plots of a_{tan} vs t, a_c vs t, and a_{tot} vs t. Include both stages in your plots.
d) Determine the maximum net force (magnitude) the train exerts on a 70.0 kg rider during the turn.
e) Contemplate a ball hanging from the ceiling of the train. Describe the motion of the ball relative to the train as it travels along the quarter-circle track. Explain how the ball is an accelerometer.

10.3 A flywheel has radius of 0.20 m. The flywheel rotates in the horizontal plane (axis of rotation is dotted line). A 15 gram mass is attached to the rim of the flywheel with superglue. A plot of angular velocity versus time for the flywheel is shown at right. Assume the initial angular positon is zero.

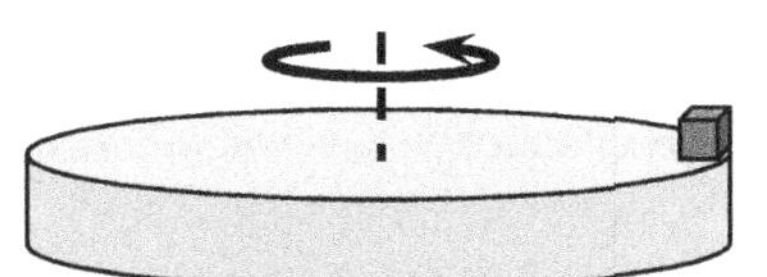

a) At what time does the mass return to its initial position?
b) How far has the mass traveled after 8 seconds?
c) Sketch a plot of α vs t.
d) How do you get a_{tan} vs t from α vs t?
e) Sketch a plot a_c vs t.
f) At what time(s), if any, is acceleration of the mass zero?
g) Determine the magnitude of the largest force experienced by the mass during the time interval shown.
h) Think quickly: how do you get RPM from this plot?

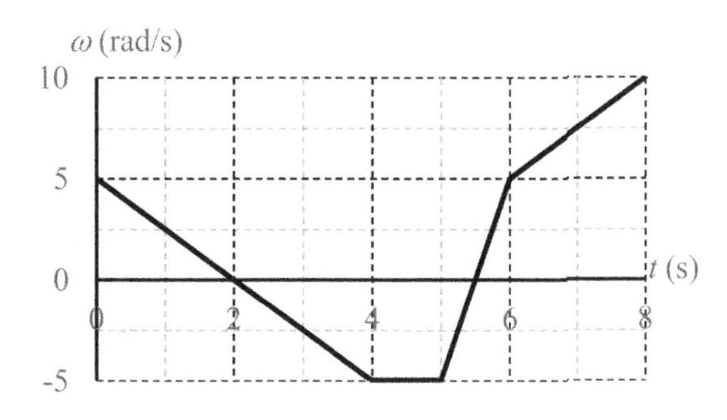

10.4 A wheel has radius $r = 0.200$ m and rotates in the xz-plane. The angular velocity of the wheel is given by $\omega(t) = 8.00t - 6.00t^2$.

a) Determine the units implied for the numbers 8.00 and 6.00.
b) According the figure at right, what unit vector is implied on $\omega(t)$?
c) Determine the number of revolutions completed by the wheel when it has maximum rotation rate in the positive direction.
d) At what time does the wheel reverse direction?
e) At what time, if any, do points on the wheel experience no net force?
f) Plot θ vs t, ω vs t, and α vs t for the first two seconds of motion. Assume initial angular position is zero.

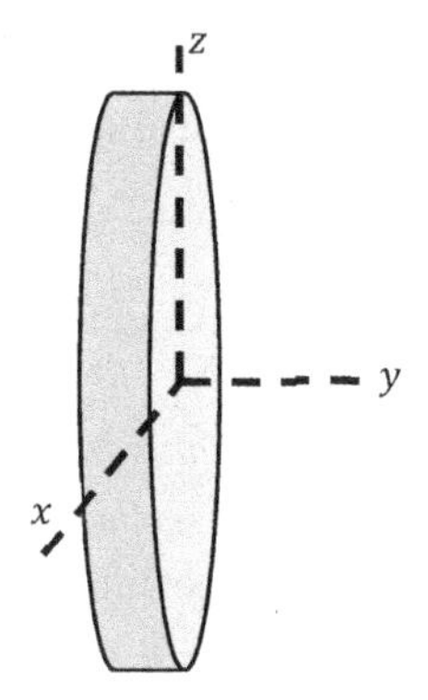

10.5 A disc with initial rotation rate ω_0 slows down with rate $\alpha = k\omega^2$ where k is a constant.

a) Determine $\omega(t)$.
b) Based on your previous experience with kinematics, what kind of situation might be described by this model? What type of force might give rise to the equation $\alpha = k\omega^2$?
c) **Feeling frisky?** Determine $\theta(t)$ and $\alpha(t)$. Sketch plots of θ vs t, ω vs t, and α vs t.

10.6 A train is at Nanzih Export Processing Zone Station (NEPZ) headed for Oil Refinery Elementary School Station (ORES). The transit between the two stations can be split into three segments. First the train accelerates from rest with constant magnitude a for distance L. Next, the train travels at constant speed for a circular bend with length L and radius $2L$. Finally, the trains slows to a stop with the same magnitude a in distance L. Kaohsiung is in the house; figure not to scale!!!

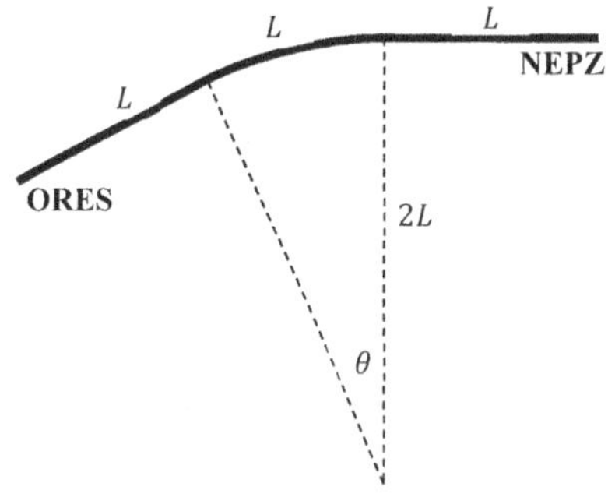

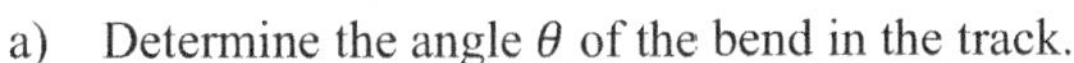

a) Determine the angle θ of the bend in the track.
b) Determine the speed during the second segment.
c) Determine the ratio of the elapsed times of the first and second segment.
d) Sketch a_{tot} vs t. Remember, $a_{tot} = \|\vec{a}_{tot}\|$.
e) Imagine a ball is hanging from the ceiling of the train. Describe the motion of the ball relative to the train during the travel. Is this a constant acceleration problem?

10.6½ The angular acceleration of a flywheel versus time is recorded using a sensor. The plot is shown at right. The flywheel is essentially a solid disk of mass 0.950 kg and radius 0.330 m rotating about an axis through the center, perpendicular to the plane of the disk (rotating $\pm\hat{k}$). At time $t = 0$ s, the flywheel rotates 48.0 RPM (in the positive direction). **I will assume the +/- sign on angular quantities indicates direction so no unit vectors are needed.**

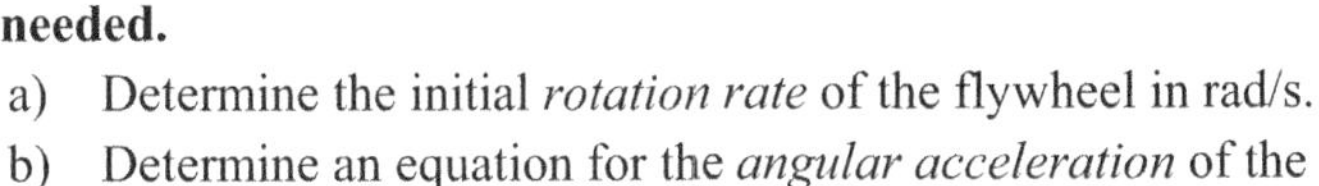

a) Determine the initial *rotation rate* of the flywheel in rad/s.
b) Determine an equation for the *angular acceleration* of the flywheel as a function of time.
c) Determine the *angular velocity* as a function of time.
d) Determine the *rotational energy* of the flywheel at $t = 8.00$ s in Joules.

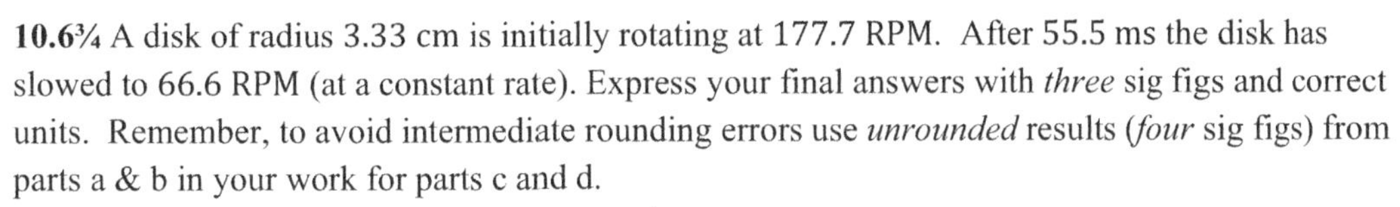

10.6¾ A disk of radius 3.33 cm is initially rotating at 177.7 RPM. After 55.5 ms the disk has slowed to 66.6 RPM (at a constant rate). Express your final answers with *three* sig figs and correct units. Remember, to avoid intermediate rounding errors use *unrounded* results (*four* sig figs) from parts a & b in your work for parts c and d.

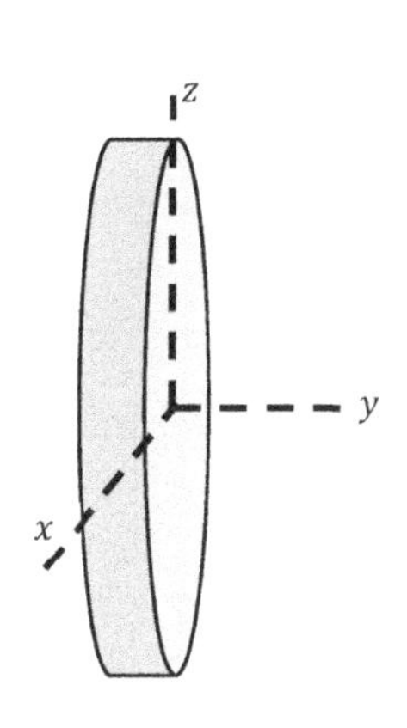

a) Convert the initial rotation rate to $\frac{\text{rad}}{\text{s}}$.
b) Determine the *magnitude* of angular acceleration (in $\frac{\text{rad}}{\text{s}^2}$).
c) How many *revolutions* does the flywheel complete over the 55.5 ms time interval?
d) When does total acceleration (magnitude) of a point on the edge of the disk equal g?

Relating Rotation to Translation

We have seen angular kinematics and rotational kinematics were closely related. While there were a few equations you just have to memorize (e.g. $a_c = r\omega^2$, $a_{tot} = \sqrt{a_c^2 + a_{tan}^2}$, or $\omega = 2\pi f = \frac{2\pi}{T}$. Fortunately, for most of the equations we could use previous results that most of us have memorized by now. We simply shifted the variables from translational to rotational.

It turns out mass, force, and momentum also have rotational equivalents. Many of our old equations apply if we correctly substitute the equivalent rotational variables. To reduce the tediousness of copying equations off the board, I tried to make a game of it.

10.7 Use the relationships in the first table to write out the corresponding formulas in the lower table.

Translational Quantity		**Rotational Quantity**	
Position	$\vec{x}$	Angular Position	$\vec{\theta}$
Velocity	$\vec{v}$	Angular Velocity	$\vec{\omega}$
Acceleration	$\vec{a}$	Angular Acceleration	$\vec{\alpha}$
Mass	m	Moment of Inertia	I
Force	$\vec{F}$	Torque	$\vec{\tau}$
Linear Momentum	$\vec{p}$	Angular Momentum	$\vec{L}$

Translational Equations		**Rotational Equation**	
Newton's 2nd Law (style 1)	$\Sigma\vec{F} = m\vec{a}$	Newton's 2nd Law (style 1)	
Newton's 2nd Law (style 2)	$\Sigma\vec{F} = \frac{d\vec{p}}{dt}$	Newton's 2nd Law (style 2)	
Work (constant force)	$W = F_x x$	Work (constant ________)	
Work (style 2)	$W = \int_i^f \vec{F} \cdot d\vec{s}$	Work (style 2)	
Kinetic Energy (TKE)	$K = \frac{1}{2}mv^2$	Kinetic Energy (RKE)	
Power	$\mathcal{P}_{inst} = \vec{F} \cdot \vec{v}$	Power	
Momentum	$\vec{p} = m\vec{v}$	Angular Momentum	

Moment of Inertia

Mass is a pretty simple concept. The greater the mass, the greater the resistance to CHANGES in motion. To get the mass of an object you slap it on a balance and you're done. If you have to calculate mass I suppose you could rearrange $\rho = \frac{m}{V}$ to give $m = \rho V$. **In contrast, moment of inertia is a world of hurt.** Sometimes I get lazy and say rotational mass in lieu of moment of inertia but in the real world this term makes no sense.

Moment of inertia describes the tendency of an object to resist changes in *rotational* motion. Moment of inertia depends on four things:

1) Total amount of mass
2) Mass distribution (shape)
3) Axis location
4) Axis orientation

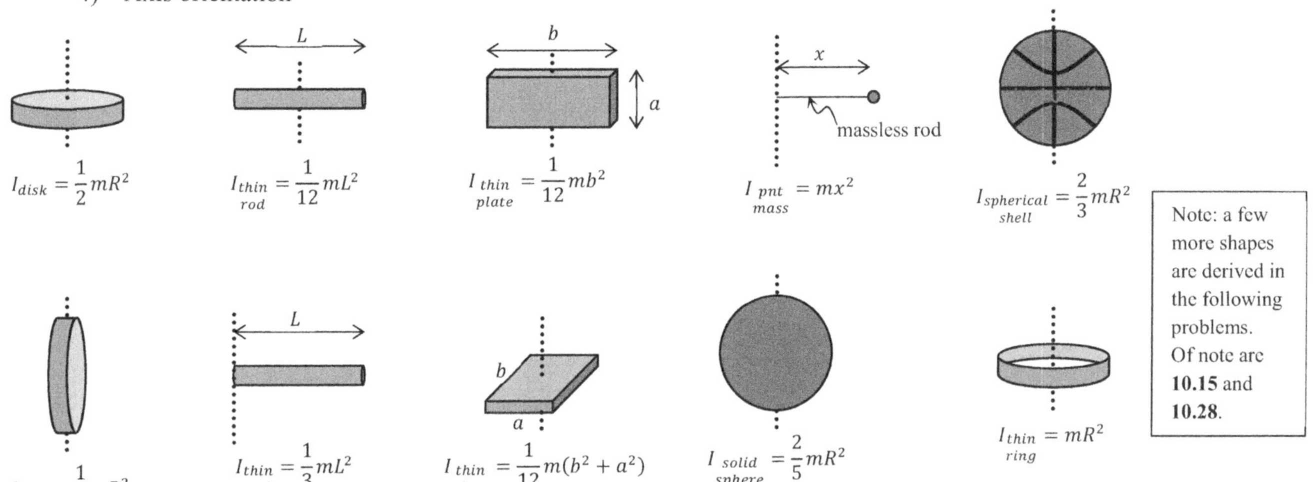

In the first column we see two different entries for identical disks. The second column shows two entries for identical rods. The third column shows two options for a thin rectangular plate. This will obviously be confusing to new physics students. Furthermore, every shape has an *infinite number* of different possible moments of inertia…

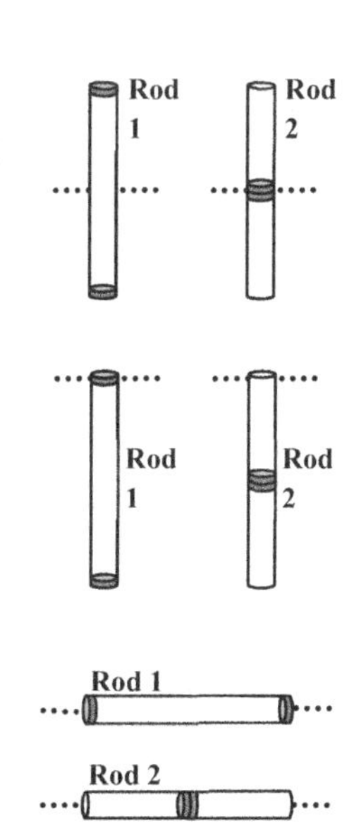

10.8 Two pipes have mass m, length L, and radius $r = \frac{L}{25}$. Each pipe/rod has two point masses (small grey cylinders) attached internally. Each point mass is $M = 5m$. Both rods are rotated back and forth using the axis shown. **For 10.8 & 10.9, treat the grey cylinders as point masses.**

a) Before calculating anything, take a guess? Which rod is easier to twist? Is it a tie?
b) Calculate I for each wand in terms of m and L. Use $I_{tot} = I_{rod} + I_{pnt\ mass\ 1} + I_{pnt\ mass\ 2}$.

10.9 The same rods as in the previous problem are now rotated using the *different axis* shown.

a) Before calculating anything, take a guess? Which rod is easier to twist? Is it a tie?
b) Now calculate the moment of inertia of each wand in terms of m and L.

10.10 Same rods as in the previous problem are now rotated using the *different axis* shown.
Watch out: in this part you must treat the grey cylinders as disks and each pipe as a ring!

a) Before calculating anything, take a guess? Which rod is easier to twist? Is it a tie?
b) Now calculate the moment of inertia of each wand in terms of m and L.
   For our wands $L = 1$ m, $r = 4$ cm, $m \approx 100$ g.
c) Tabulate the results of **10.8-10.10** using decimals (not fractions). This makes it easier to compare.

Using Calculus to Determine Moment of Inertia

The definition of moment of inertia for continuous mass distributions is

$$I = \int r^2\, dm$$

where r is distance from the axis of rotation to a slice and dm is the mass of a slice. An example of this is shown at right for a sphere. **Notice all points in the cylindrical slice are equidistant from the axis.**

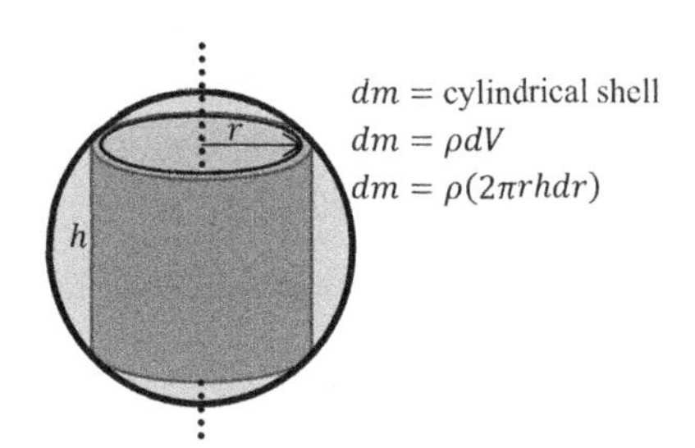

Side note: you could use triple integrals with cylindrical symmetry for any 3D object. This requires slicing the object into point masses using $dm = \rho(rdrd\theta dz)$ and integrating over total volume. This is beyond the scope of this course.

In practice, many of the intimidating integrals simplify dramatically as shown in the examples below.

Shape	Integral	Figure
Thin rod	$I = \int_i^f x^2\, dm$	x; $dm =$ point mass; $dm = \lambda dx$
Thin plate	$I = \int_i^f x^2\, dm$	x; $dm =$ rectangular strip; $dm = \sigma dA$; $dm = \sigma(hdx)$
Thin plate with cylindrical symmetry	$I = \int_{R_i}^{R_o} r^2\, dm$	r; $dm =$ circular strip; $dm = \sigma dA$; $dm = \sigma(2\pi rdr)$; **Axis thru center runs out of page**

When doing calculus problems, four main parameters vary during practical computations:

1) Location of the axis
2) Size of the object
3) Density of the material (uniform vs non-uniform)
4) Axis orientation

The first two parameters affect the limits of integration while the third affects the integrand. The fourth parameter, axis orientation can often be handled with a cool math trick known as the perpendicular axis theorem. Let's do some examples…

10.11 2D Calculus Uniform density A rectangular plate has uniform mass density. The plate has length a and height b. The total mass of the plate is M. The plate is to be rotated about the y-axis shown as a dashed line.

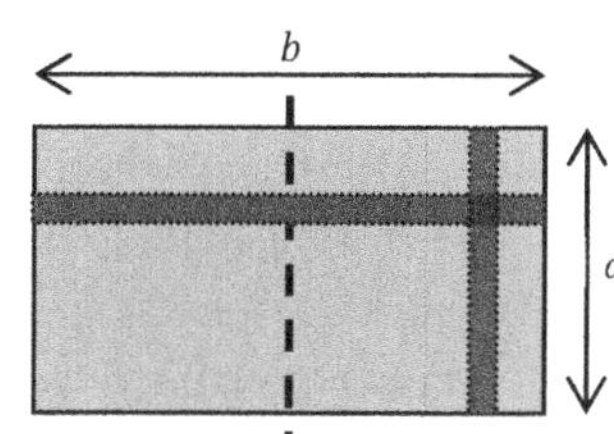

a) Which of these two slices gives the same r for every part of the slice? That is the correct slice to determine I_{yy} (the moment of inertia about the y-axis).
b) Determine the mass density in terms of M, a, and b.
c) Determine the differential mass of the slice dm.
d) Determine the moment of inertia in terms of a, b, and M. Compare to the table.

10.12 2D Calculus Uniform density The same uniform rectangular plate as the previous problem is used with a different axis of rotation. In the previous problem the center of mass used the integral

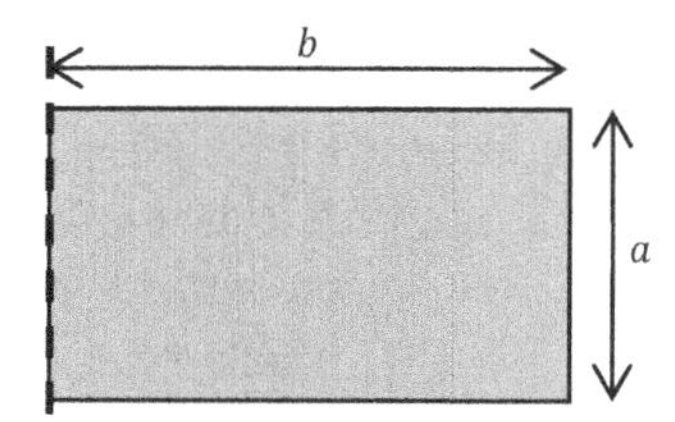

$$I_{yy} = \int_{-b/2}^{+b/2} \sigma a x^2 dx$$

How does this integral change if the new axis is used instead?

10.13 2D Calculus uniform density – Result used in 10.19, 10.23, & 10.30-31
A uniform density triangular plate of mass m is to be rotated with the axis shown.

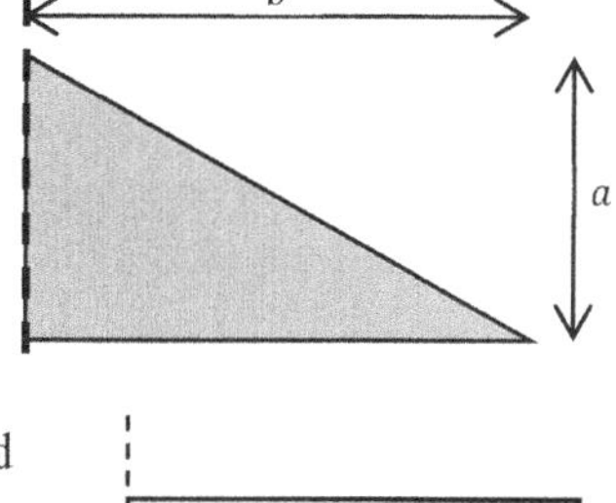

a) Calculate the moment of inertia about the axis shown.
b) Suppose this triangle was made of the same material as the rectangular plate. Should the moment of inertia be half, slightly more than half, or slightly less than half the entire rectangle of problem **10.12**?

10.14 2D, Calculus, Non-uniform density: A thin rod of length L has density $\lambda = \alpha x^2$ based on the coordinate system shown. The total mass of the rod is M but α is an unknown constant. The rod is rotated about the y-axis shown as a dotted line in the figure.

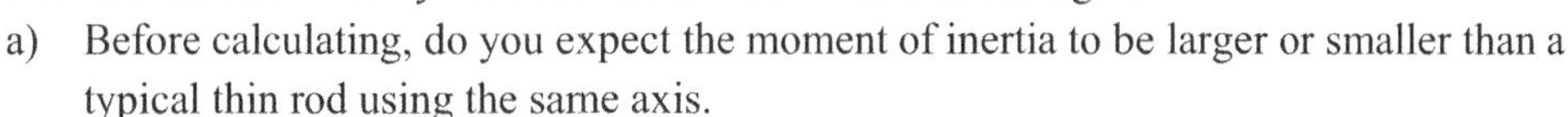

a) Before calculating, do you expect the moment of inertia to be larger or smaller than a typical thin rod using the same axis.
b) Total mass is given by $M = \int dm$. Use the total mass equation to determine α in terms of M and L.
c) Determine the moment of inertia about the y-axis in terms of M and total length L. **Compare to part a.**
d) In non-uniform problems you don't get to choose your coordinate system.
If the origin in the figure was at the middle of the rod, what would change in the computation of I?

Note: strictly speaking you *could* change your coordinate system as long as you also determine how the coordinate shift alters the density function. Typically it is easier to use the provided coordinate system.

10.15 2D Calculus, Uniform density, Cylindrical symmetry: A disk of metal has a concentric hole cut from it. The radius of the disk is R_2 while the radius of the hole is R_1. The axis of rotation runs through the center of the hole pointing out of the page. This shape is called an annulus or thick ring. The thick ring has mass M.

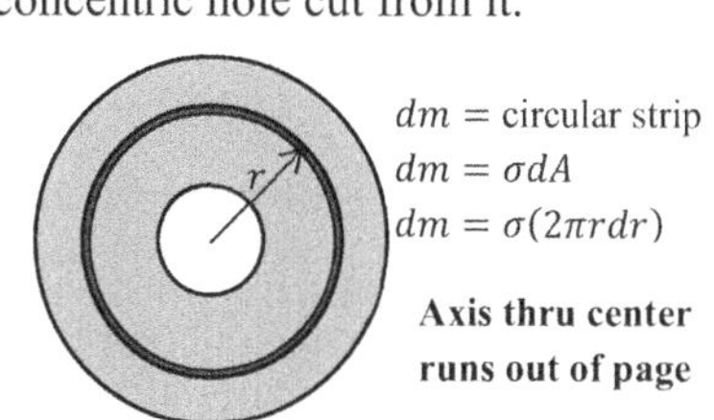

a) Determine the mass density σ.
b) Determine the moment of inertia for this shape (sometimes called an annulus or thick ring). Write your answer in terms of the givens (do not leave it in terms of σ).

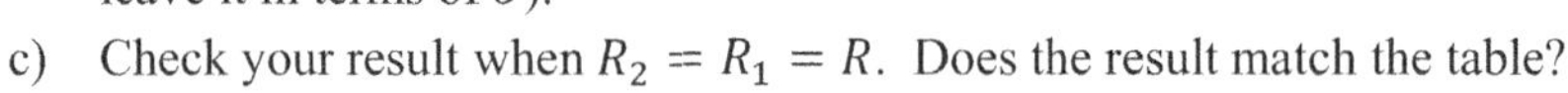

c) Check your result when $R_2 = R_1 = R$. Does the result match the table?
d) Check your result when $R_2 = R$ and $R_1 = 0$. Does the result match the table?

Parallel Axis Theorem

To appropriately apply this theorem, two conditions must apply:

1) The two axes are *parallel* to each other.
2) One of the axes runs through the *center of mass* (indicated by the black dot).

When both conditions are met, we find that

$$I_{\parallel axis} = I_{\text{CM}} + md^2$$

where I_{CM} is the moment of inertia using the axis running through the center of mass and d is the distance between the two parallel axes.

10.16 Consider a uniform rod with mass M and length L. The dashed axis runs through rod's center. An axis of rotation is indicated by the dashed line. It runs through a point distance x from the left end of the rod.

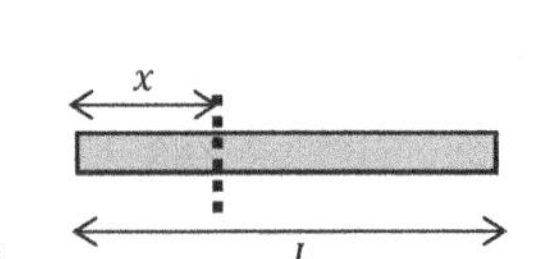

a) Use the parallel axis theorem to determine I about the axis shown. Before you check your answer, do the next two parts…
b) Think: what should your result be when $x = 0$? Compare your result to the table.
c) What should the result be if $x = \frac{L}{2}$? Does your answer for part a) give correct results for parts b) and c)?

10.17 Consider a uniform sphere of mass M and radius R. An unhappy face is painted on the sphere for no good reason. What is the moment of inertia if someone rotates the sphere using the dotted axis?

10.18 A bike wheel hangs on a nail as shown. The wheel is made from a thin ring of mass M and radius R and three spokes each of mass M and length R. Determine the moment of inertia of the wheel assuming the nail is the axis of rotation.

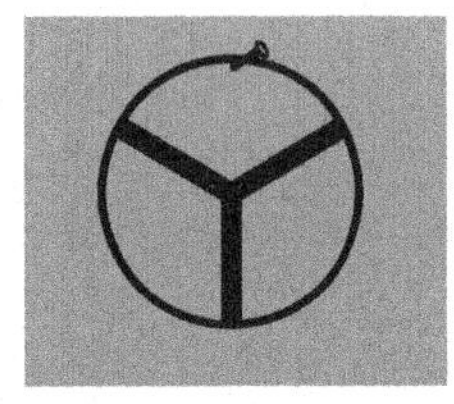

10.19 Dracula takes a break from sucking the blood. He sees the triangular plate with mass m and side L as shown. Using the dotted line axis, the triangular plate has moment of inertia

$$\frac{1}{6}mL^2 \quad \text{(from problem } \mathbf{10.13a}\text{)}$$

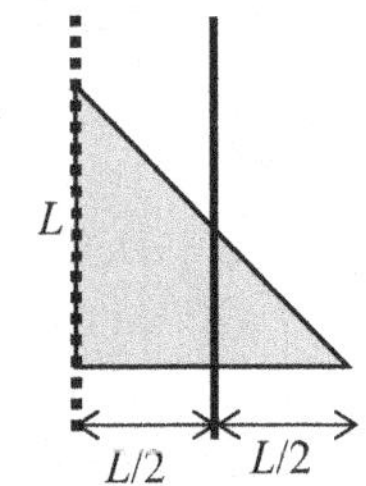

Dracula telepathically communicates to you the following computation and associated trash talk:

"I do not need zee zilly calculus to determine the moment of inertia for the solid line axis. I vill use the parallel axis theorem like this:

$$I_{dotted} = I_{solid} + m\left(\frac{L}{2}\right)^2$$

It's so easy!"

Notice that Dracula's proof gives a negative answer for the moment of inertia…clearly wrong. Your physics teacher, hopped up on garlic and clearly getting loopy from writing down one too many moment of inertia problems, says, "Go Bulldogs! My students will tell you the error of that foolish maneuver…I hope!"

a) Why does his method not work? What did he do wrong or forget about?
b) Derive the correct answer of $I_{solid} = \frac{1}{12}mL^2$.

10.20 Frankenstein decides to get in on the mix. "Dude," he says, "Consider the uniform disk shown at right in the figure. From the table $I_{\text{CM}} = \frac{1}{2}MR^2$. Using the axis shown in one finds $I = \frac{3}{2}MR^2$. My physics teacher says this is total crap but whatever…"

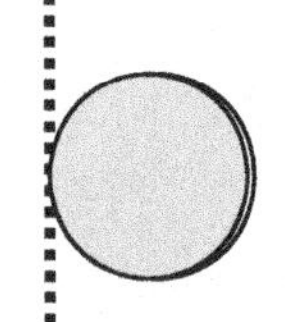

a) Why is Frankenstein's application of the parallel-axis theorem inappropriate?
b) Show $I_{dotted} = \frac{5}{4}MR^2$.

Perpendicular Axis Theorem

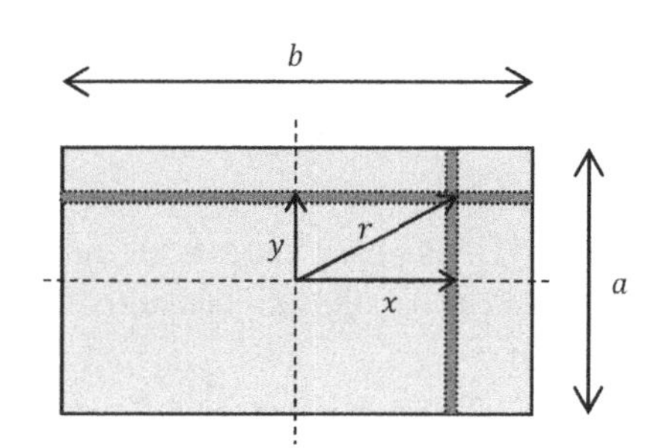

The figure at right shows a rectangular plate in the xy-plane with mass m and sides a & b. To determine the moment of inertia about the x-axis (I_{xx}) we use the horizontal slice and obtain $I_{xx} = \frac{1}{12}ma^2$. Similarly $I_{yy} = \frac{1}{12}mb^2$ using the vertical slice.

But what about I_{zz}, the moment of inertia about the z-axis out of the page? From the definition of moment of inertia we find

$$I_{zz} = \int r^2\, dm$$

$$I_{zz} = \int (x^2 + y^2) dm$$

$$I_{zz} = \int x^2\, dm + \int y^2\, dm$$

$$I_{zz} = I_{xx} + I_{yy}$$

This last result is called the perpendicular axis theorem. **This result is only valid for 2D objects in the xy-plane.** A quick way to see this is to try to use the perpendicular axis theorem on a solid sphere…it doesn't work!

10.21 Use the perpendicular axis theorem to determine I_{zz} for the rectangular plate shown above. Compare your result to the table provided earlier.

10.22 A metal hoop has mass M and radius R.

a) Determine the moment of inertia of metal hoop about axis **1**.
b) Determine the moment of inertia of metal hoop about axis **2**.

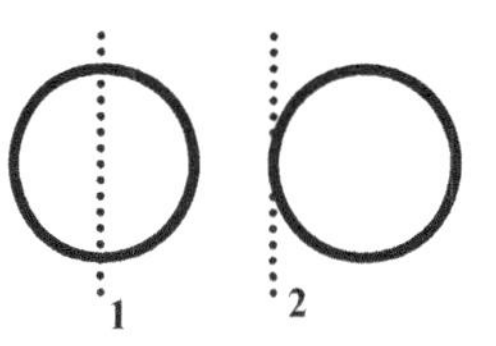

10.23 Challenge: A thin plate is shaped like an equilateral triangle of mass m and side s. An axis of rotation runs through the center of mass and points out of the page shown by the **x** in the figure. Determine the moment of inertia. Hint: use the result of **10.13a**. With this result, no calculus is needed.

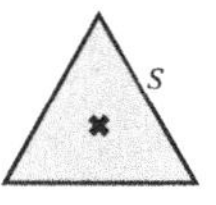

Summary on how to calculate the moment of inertia:

1) Identify the axis location and orientation.
2) Hopefully you can split the object into parts from the table and sum the moment of inertias using
$$I_{tot} = I_1 + I_2 + \cdots$$
3) If an object and axis are similar to the table but the axis is shifted *parallel* to the one shown in the table, use the parallel axis theorem:
$$I_{\parallel axis} = I_{\text{CM}} + md^2$$
Remember that the following two criteria must be met to use the parallel axis theorem:
    - The two axes are *parallel* to each other.
    - One of the axes runs through the *center of mass*.
4) If a *planar* object (2D plate) and axis are similar to the table but the axis is shifted *perpendicular* to the one shown in the table, use the perpendicular axis theorem:
$$I_{zz} = I_{xx} + I_{yy}$$
We can use this equation for 2D objects that lie in the xy-plane.
5) One can always use calculus and the basic definition
$$I = \int r^2 dm$$
For non-uniform density this is a must.

Some general info to give intuition about moment of inertia problems:

- When most of the mass is close to the axis, I is small
- Small I implies easy to *CHANGE* rotation
- For any chosen axis orientation, I_{CM} is always smaller than any $I_{\parallel}$
- For *planar* objects, moment of inertia about axis out-of-plane (I_{zz}) is always ≥ moments of inertia in-plane

10.23½ REVIEW A square plate of side s and density σ has a circular hole cut out of it as shown. Notice the diameter of the hole is *approximately* $\frac{s}{2}$. For our purposes, you may assume it is *exactly* $\frac{s}{2}$ even though such a hole, in real life, would cause the bottom right corner of the square to separate from the plate!

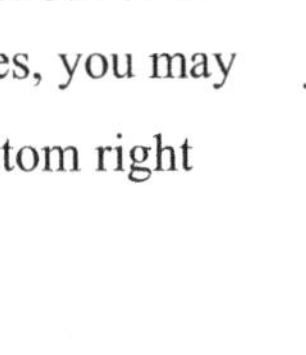

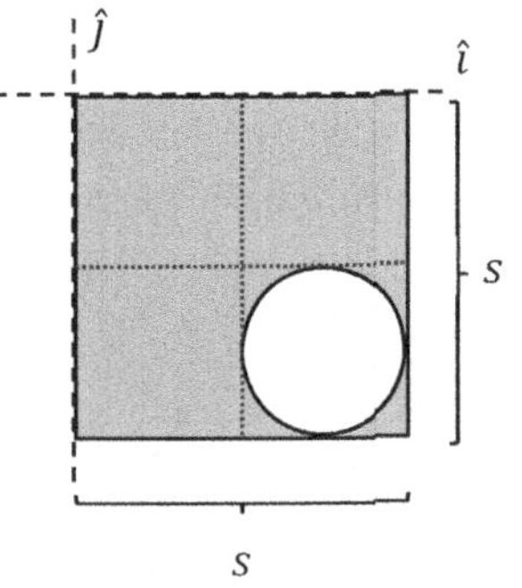

a) Does the positive z-axis point into or out of the page?
b) Based on the coordinates shown, determine the center of mass location of the plate (after the hole is cut).
c) Determine the moment of inertia if the plate is rotated about the x-axis (I_{xx}).
d) Determine the moment of inertia if the plate is rotated about the y-axis (I_{yy}). Is it the same as the result from part c?
e) Determine the moment of inertia if the plate is rotated about the z-axis (I_{zz}). Is it the same as the result from part c?

10.23¾ See if you can get these WITHOUT computation (Concept Q). What if the thin plate from the previous problem is reoriented such that the hole is now located as shown in the figure at right. How, if at all, do any of the following quantities change?

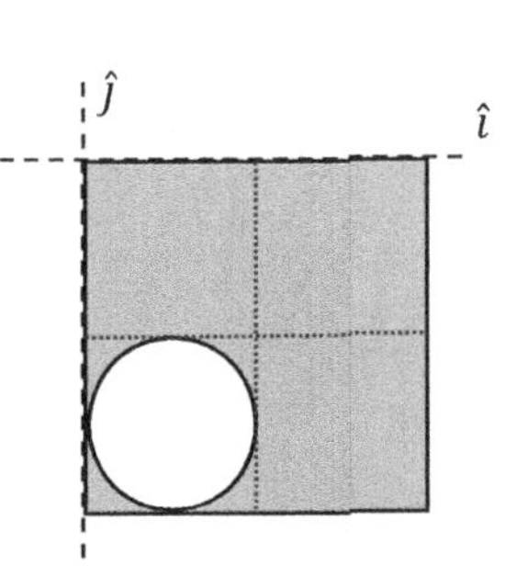

a) x_{CM}
b) y_{CM}
c) I_{xx}
d) I_{yy}
e) I_{zz}

10.24 Choking up on a baseball bat implies you move your hands away from the end of the bat and closer to the center. How does choking up on a baseball bat affect the moment of inertia during a typical swing? Explain.

10.25 1D Calculus A thin tube of gas with length $L = 2.50$ m and mass $M = 0.120$ kg has non-uniform density

$$\lambda = \alpha x e^{-2.00x^2}$$

based on the coordinate system shown. The rod is to be rotated about the y-axis. Assume the rod is long enough that we may use $L \approx \infty$. An integral table is provided below.

a) What units are implied on the number 2.00?
b) Determine the constant α to three sig figs.
c) Determine I_{yy} (the moment of inertia about the y-axis).

Note: a similar type of integral appears in quantum mechanics as the density of electron clouds. The process of determining the constant α is similar to a crucial step in a process known as normalizing the wavefunction.

$I_0 = \int_0^\infty e^{-ax^2}dx = \frac{1}{2}\sqrt{\frac{\pi}{a}}$	$I_1 = \int_0^\infty xe^{-ax^2}dx = \frac{1}{2a}$
$I_2 = \int_0^\infty x^2e^{-ax^2}dx = -\frac{dI_0}{da} = \frac{1}{4}\sqrt{\frac{\pi}{a^3}}$	$I_3 = \int_0^\infty x^3e^{-ax^2}dx = -\frac{dI_1}{da} = \frac{1}{2a^2}$
$I_4 = \int_0^\infty x^4e^{-ax^2}dx = \frac{d^2I_0}{da^2} = \frac{3}{8}\sqrt{\frac{\pi}{a^5}}$	$I_5 = \int_0^\infty x^5e^{-ax^2}dx = \frac{d^2I_1}{da^2} = \frac{1}{a^3}$
$I_{2n} = \int_0^\infty x^{2n}e^{-ax^2}dx = (-1)^n\frac{d^nI_0}{da^n}$	$I_{2n+1} = \int_0^\infty x^{2n+1}e^{-ax^2}dx = (-1)^n\frac{d^nI_1}{da^n}$

10.26 Two thin square plates are made of the same material. The larger has side s. Both plates are designed to be rotated with axis of rotation along the side of the square. The moment of inertia of the smaller square is half of the larger square. Determine the side of the smaller square in terms of s. The answer surprised me…

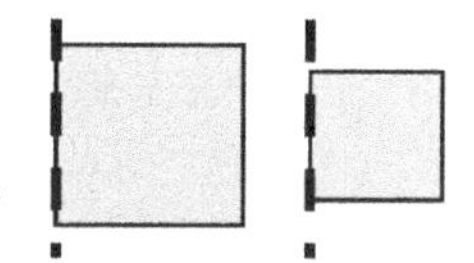

10.27 A thin rod of length L has non-uniform density given by $\lambda = \alpha x$ based on the coordinate system shown. The rod is to be rotated about the y-axis shown.

a) Based on the density function, where do you expect the center of mass to be: in the middle, towards the right end, or towards the left end?
b) Determine the center of mass of the non-uniform object as a simplified fraction times L.
c) Determine the total mass of the object in terms of α and L.
d) Determine I_{yy} (moment of inertia about the y-axis) in terms of the total mass M and total length L. Think: do you expect it to be more or less than $\frac{1}{3}ML^2$?

10.28 A wheel of mass M is made from a thin ring of radius R and three large spokes as shown in the figure. The owner of wheel is curious how much of the mass comes from the spokes versus the ring. Using an axis through the center of mass that runs out of the page, the woman determines the moment of inertia is I. Determine the mass of each spoke and the mass of the wheel in terms of M, R, and I. To be clear, M is the total mass of the rim and all three spokes.

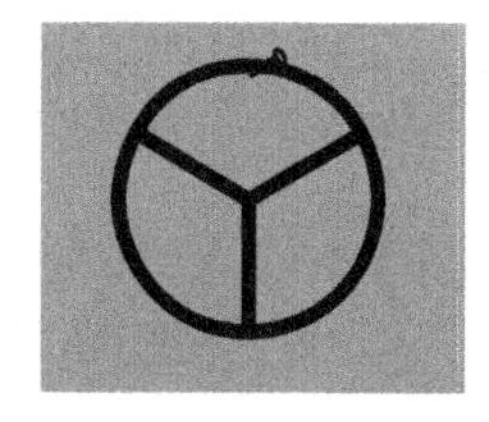

10.29 I suppose in real life it would be easier to determine the mass M of the entire object and the moment of inertia of the object *about a point on the rim*. The inertia about a point on the rim can be found experimentally by hanging it on a nail and timing the period of small oscillations (see figure at right). We will learn more about this technique in a chapter on oscillations in another class. Try redoing the previous problem assuming you know I_{rim} in lieu of I_{CM}.

10.30 A square plate of side $L = 0.400$ m has moment of inertia $0.100\ \text{kg} \cdot \text{m}^2$ when rotated about an axis aligned with the edge of the plate as shown in the figure.

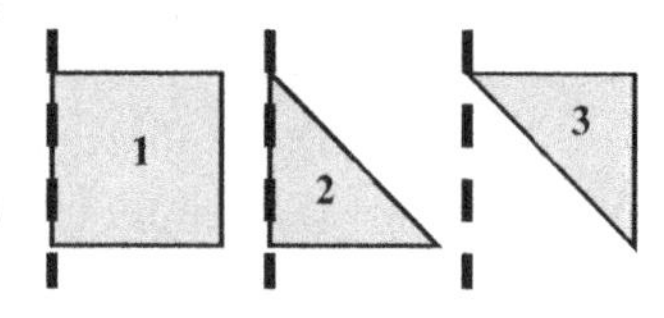

a) The plate is cut into two equal triangular pieces. Each piece will be welded to a metal rod and rotated about the axis shown. Before calculations, which triangle has the larger moment of inertia? Is it a tie?
b) Determine the moment of inertia of each triangle. Result of **10.13a** may be useful….

10.31 Five thin plates are made from aluminum. In contrast to the previous problem, this time each plate has the same total area and thus the same total mass. Rank the moments of inertia from smallest to largest; indicate any ties.

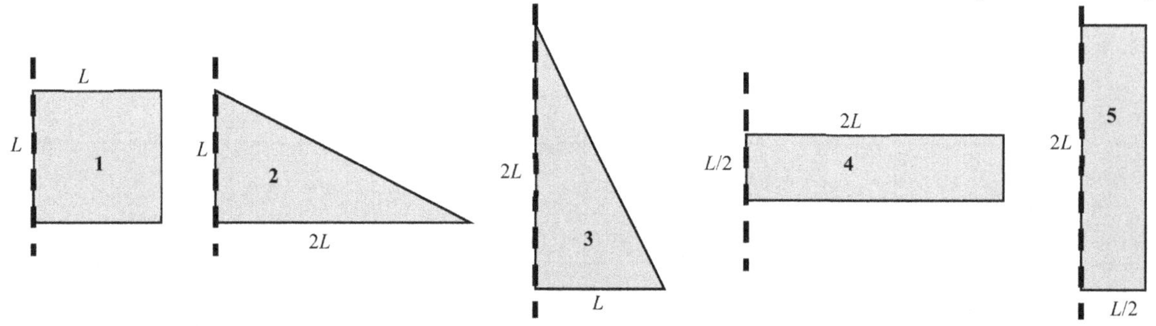

10.32 When is it appropriate to consider sphere as a point mass?
Suppose you have a sphere with mass M and radius R. The sphere can be located at any distance x along a massless rod which is free to pivot about its left end.

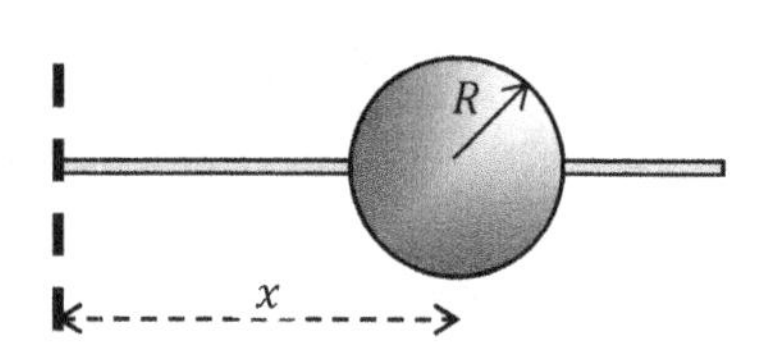

a) Determine the moment of inertia of the object in terms of M, x, and R. Do not assume the sphere is a point mass. Use the parallel axis theorem.
b) If the sphere is instead treated as a point mass the moment of inertia is the simpler formula Mx^2. For what minimum value of x is this simpler formula 99% of the moment of inertia given by part a?
c) Assume $M = 0.500$ kg and $r = 4.0$ cm. Plot I vs x for values of x between 0 and 20.0 cm.

10.33 A thin disk of radius R has an off-axis hole of radius r. The axis runs out of the page through the center of the disk shown by the **x** in the figure. The center of the hole is distance x from the axis. The disk is made of material with density σ.

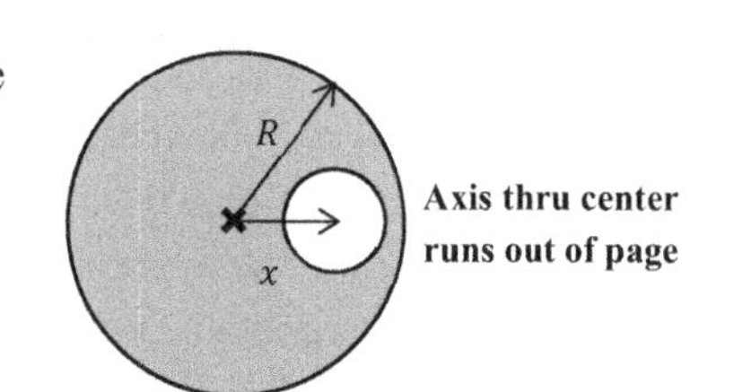

a) Determine the moment of inertia.
b) Think, how could you check your result against the result of problem **10.15**?

10.34 Three thin plates have equal mass and density. The first is a disk, the second a square, and the third an equilateral triangle. All three plates lie in the xy-plane. Each plate has an axis of rotation through its center of mass pointing out of the plane. Rank the moments of inertia from smallest to largest; indicate any ties. Note: we have already derived equilateral triangle moment of inertia (for this particular axis & orientation) as $I_{triangle} = \frac{1}{12}Ms^2$.

A Few Final Tricks for Deriving Moments of Inertia

The moment of inertia for a shape can be derived from the equation

$$I = \int_i^f dI$$

where dI is the moment of inertia of a small slice of your shape. While this equation seems obvious enough, it helps to see some examples.

For instance, consider the sphere of radius R and mass m sliced as shown at right. The density of the sphere is

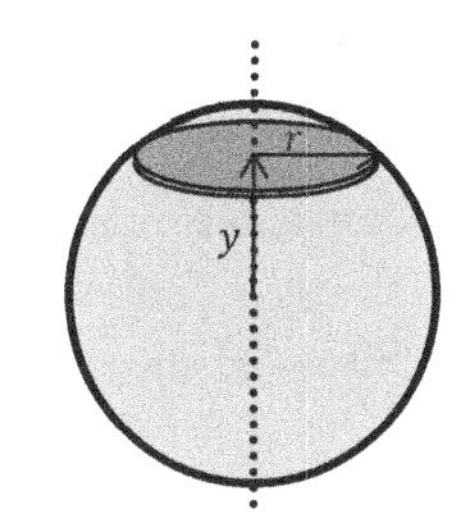

$$\rho = \frac{m_{tot}}{V_{tot}} = \frac{m}{\frac{4}{3}\pi R^3}$$

The mass of the slice is

$$dm = \rho dV = \rho(\pi r^2 dy)$$

Since the slice is a disk, the moment of inertia of the slice is

$$dI = \frac{1}{2} r^2 dm$$

From the geometry of the sphere we see

$$r^2 = R^2 - y^2$$

Putting all the pieces together gives

$$I = \int_i^f dI$$

$$I = \int_i^f \frac{1}{2} r^2 dm$$

$$I = \int_i^f \frac{1}{2} r^2 \rho(\pi r^2 dy)$$

$$I = \frac{1}{2}\pi\rho \int_i^f r^4 dy$$

$$I = \frac{1}{2}\pi\rho \int_{-R}^{R} (R^2 - y^2)^2 dy$$

$$I = \frac{1}{2}\pi\rho \int_{-R}^{R} (R^4 - 2R^2y^2 + y^4) dy$$

$$\boldsymbol{I = \pi\rho \int_0^R (R^4 - 2R^2y^2 + y^4) dy}$$

$$I = \frac{3m}{4R^3} \int_0^R (R^4 - 2R^2y^2 + y^4) dy$$

$$I = \frac{3m}{4R^3} \left[R^4 y - \frac{2R^2y^3}{3} + \frac{y^5}{5} \right]_0^R$$

$$I = \frac{3m}{4R^3} \left(R^5 - \frac{2R^5}{3} + \frac{R^5}{5} \right)$$

$$I = \frac{3mR^2}{4} \left(1 - \frac{2}{3} + \frac{1}{5} \right)$$

$$I = \frac{2}{5} mR^2$$

For the line in bold I used the fact $\int_{-L}^{L} f(x)dx = 2\int_0^L f(x)dx$ if $f(x)$ is an even function.

It is useful to show a variation on this technique where we slice the sphere another way. Notice the sphere is still sliced like a disk. We still find

$$\rho = \frac{m_{tot}}{V_{tot}} = \frac{m}{\frac{4}{3}\pi R^3}$$

The mass of the slice is now

$$dm = \rho dV = \rho(\pi r^2 dx)$$

while

$$r^2 = R^2 - x^2$$

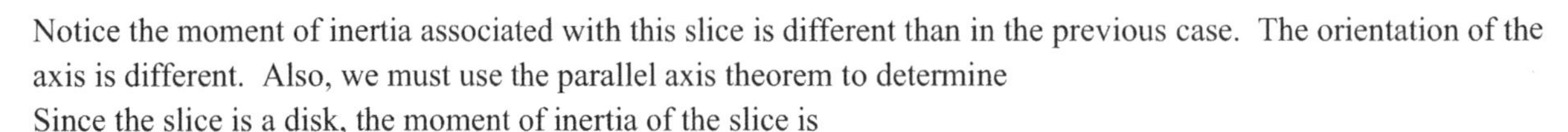

Notice the moment of inertia associated with this slice is different than in the previous case. The orientation of the axis is different. Also, we must use the parallel axis theorem to determine
Since the slice is a disk, the moment of inertia of the slice is

$$dI_{\parallel} = dI_{\mathrm{CM}} + x^2 dm$$

$$dI_{\parallel} = \frac{1}{4}r^2 dm + x^2 dm$$

$$dI_{\parallel} = \frac{1}{4}(R^2 - x^2)dm + x^2 dm$$

$$dI_{\parallel} = \frac{1}{4}(R^2 + 3x^2)dm$$

To determine the moment of inertia of the sphere you once again plug in the pieces and turn the crank

$$I = \int_i^f dI_{\parallel}$$

$$I = \int_i^f \frac{1}{4}(R^2 + 3x^2)dm$$

$$I = \frac{1}{4}\int_i^f (R^2 + 3x^2)\rho(\pi r^2 dx)$$

$$I = \frac{\pi\rho}{4}\int_{-R}^{R} (R^2 + 3x^2)(R^2 - x^2)dx$$

$$I = \frac{\pi\rho}{2}\int_0^R (R^4 + 3R^2x^2 - R^2x^2 - 3x^4)dx$$

$$I = \frac{3}{8}\frac{m}{R^3}\int_0^R (R^4 + 2R^2x^2 - 3x^4)dx$$

$$I = \frac{3}{8}mR^2\left(1 + \frac{2}{3} - \frac{3}{5}\right)$$

$$I = \frac{2}{5}mR^2$$

10.35 Until now, we have dealt with thin rods where $R \ll L$. Now consider the thick rod shown at right to be rotated about the axis shown through the center of mass. Use our new technique to derive

$$I_{\substack{thick\\rod}} = \frac{1}{4}mR^2 + \frac{1}{12}mL^2$$

Challenge: Determine the moment of inertia of a cube of mass m and side s with the axis along the main diagonal. I believe I read somewhere the answer is $\frac{1}{6}ms^2$ but I haven't verified. This should keep you busy for a while…At some point you'll almost certainly get stuck and look online. When you do you will discover there is another style, beyond the scope of this course, using matrix methods. Moment of inertia is surprisingly interesting, a lot cooler than its translational equivalent (mass).

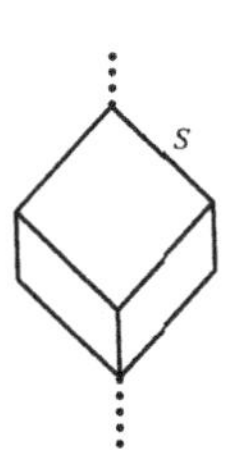

TORQUE

The methods to find torque for this class are:

1) $\vec{\tau} = \vec{r} \times \vec{F}$ (use unit vectors and wheel of pain)
2) $\vec{\tau} = rF \sin\theta$ (use *top* figure at right)
3) $\vec{\tau} = r_{\perp}F$ (use *middle* figure at right)
4) $\vec{\tau} = rF_{\perp}$ (use *bottom* figure at right)

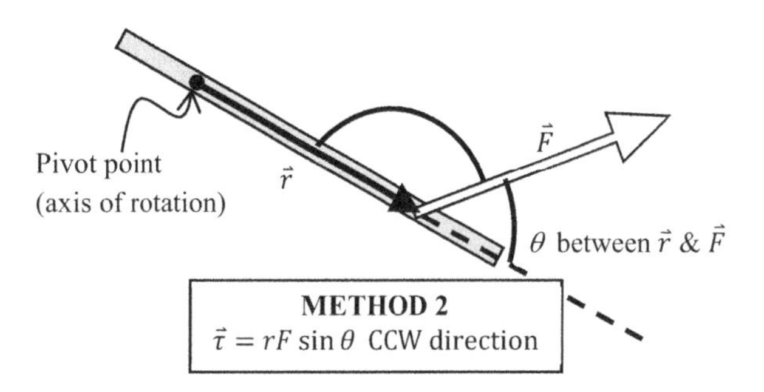

$\vec{\tau}$ = torque (units are N·m)
$\vec{F}$ is an applied force
$\vec{r}$ is the displacement **from** pivot **to** point force applied
Line of action is a line drawn through the force that extends to infinity in both directions.
$r_{\perp}$= the lever arm = $r \sin\theta$ = the perpendicular distance from the line of action to the pivot point
$F_{\perp}$= the component of $\vec{F}$ perpendicular to $\vec{r}$

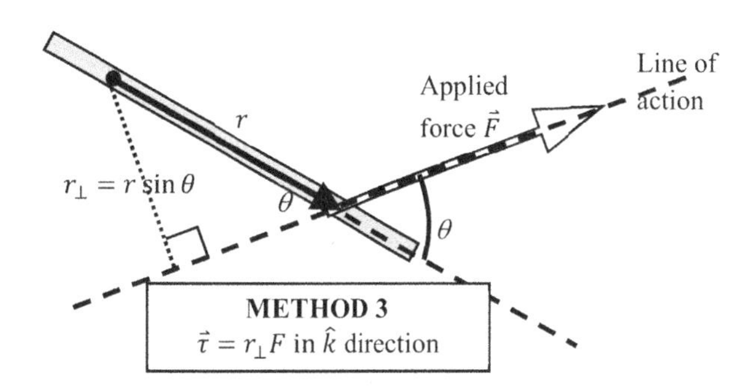

Stuff to memorize:

- If the line of action for $\vec{F}$ goes through the pivot point, $\vec{F}$ causes no torque.
- $\sin(90 - \theta) = \cos\theta$
- $\sin(180 - \theta) = \sin\theta$
- $\Sigma\vec{\tau} = I\vec{\alpha}$
- $\vec{\tau}$ is a vector, remember the direction.
- In standard math notation
  - Counter Clockwise $= +\hat{k}$
  - Clockwise $= -\hat{k}$
- Physicists often reassign the positive direction as CW if it simplifies the minus signs.
- $\vec{r}$ is the displacement **from** pivot **to** point where force is applied

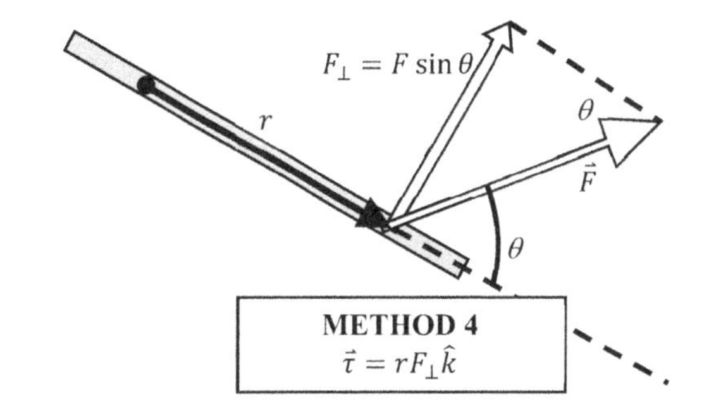

Right hand rule for determining the direction of a torque:

1) Point the fingers of your right hand in the direction of $\vec{r}$ (from the pivot towards an applied force).
2) Curl your fingers towards the applied force $\vec{F}$. Since there are four possible ways to point the fingers on your right hand you'll have to try out various right hand positions and see which one works.
3) The thumb on your right hand points in the direction of the torque.

Occasionally Useful Style

Split force into components then do lever arms (method 3) *on each component*. Consider the diagram at right where force $\vec{F}$ is applied to a rectangular plate with sides x and y as shown. **Notice the tail of each component is placed at the location where $\vec{F}$ is applied.** If we assume the pivot point is the black dot in the bottom right corner of the plate

$$\vec{\tau} = (xF \sin\theta - yF \cos\theta)\hat{k}$$

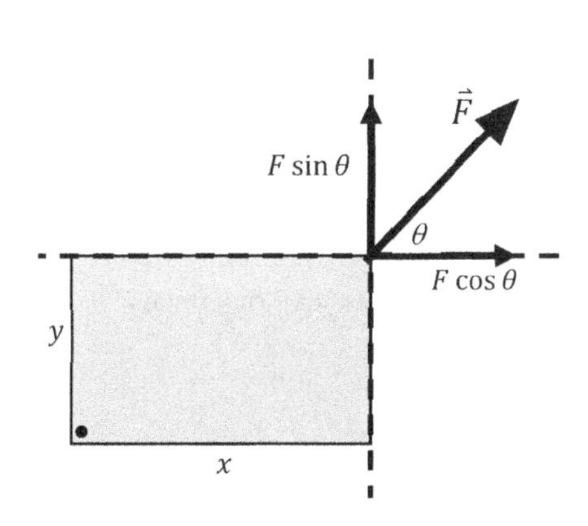

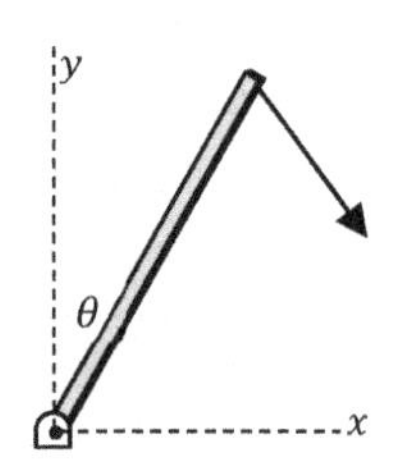

10.36 Suppose a force is $\vec{F} = (3.0\hat{\imath} - 4.0\hat{\jmath})$N is applied on a rod. The rod is pivoted about the base as shown in the figure. The angle is 30.0° from the vertical and the rod is 1.50 m long.

a) Determine $\vec{r}$ for the applied force $\vec{F}$.
b) Determine the torque caused by $\vec{F}$. Use method 1.
c) **NEW** Determine the torque caused by $\vec{F}$ by determining lever arms *for each component.*

10.37 Find the <u>direction</u> of the torque caused by $\vec{F}$ in each of the following examples. The coordinate system shown between the first two cases applies for all cases. State your answer as a unit vector.

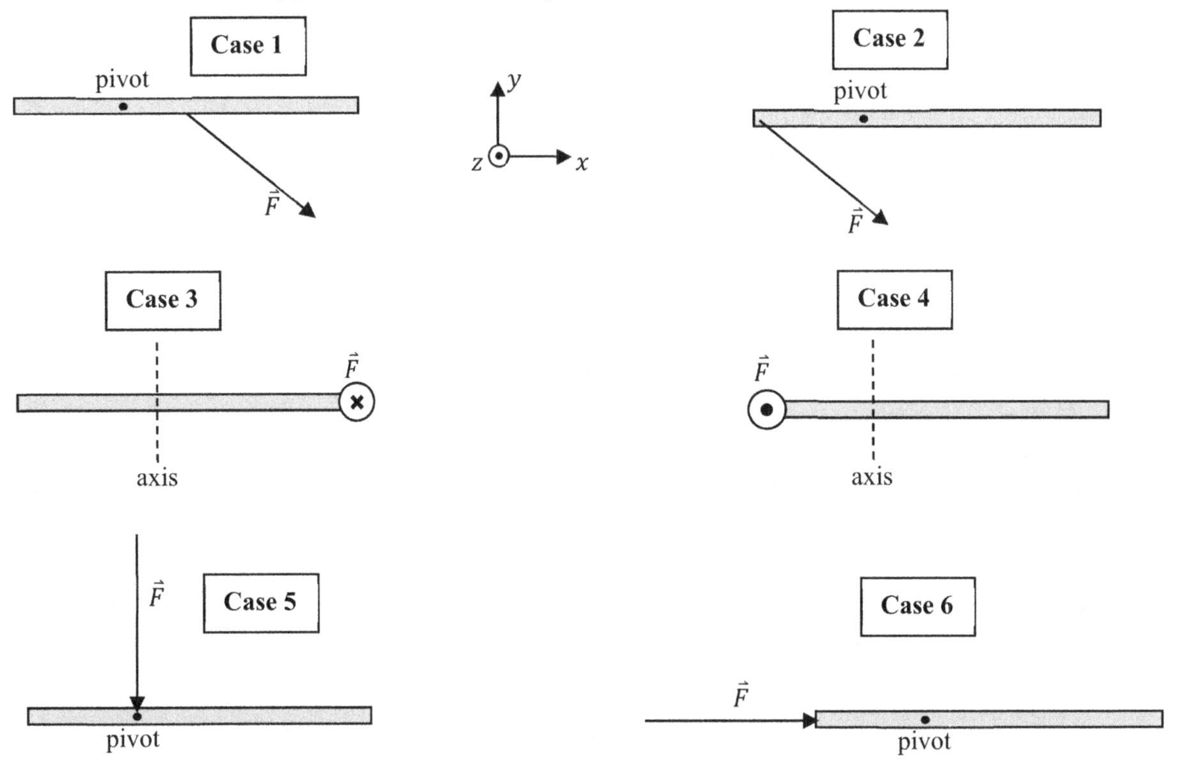

10.38 Six external forces act on a rod of length L *with negligible mass*. The rod is attached to a frictionless pivot at its center. The magnitude of each force is shown in the figure as F, $2F$, or $\frac{F}{2}$.

a) Which force(s) cause no torque?
b) Rank the <u>magnitudes</u> of the torques caused by each force from highest to lowest.
c) Which force(s) exert torques in the $+\hat{k}$ direction?
d) What is the magnitude and direction of the force exerted by the pivot to prevent the rod from *translating*? Remember, the six forces drawn in the figure are from external sources. The pivot must exert a force to balance these external forces.

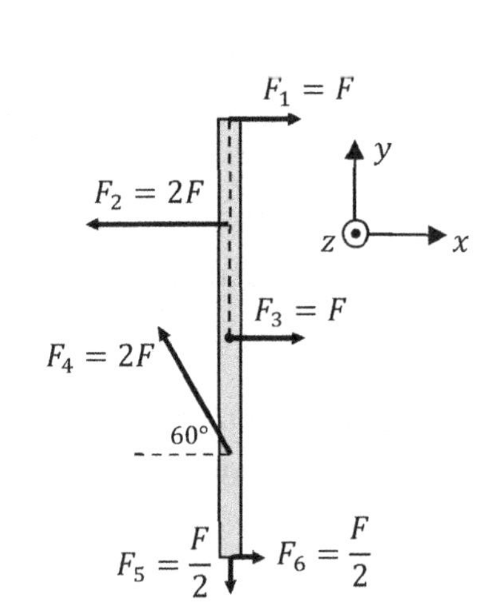

10.39 A uniform thin rod rotates in a vertical circle with a frictionless pivot at the left end. At the instant shown three forces act on the rod: the rod's weight, and applied force $\vec{F}$, and a reaction force at the pivot. The rod has mass m and length L. Determine the angular acceleration of the rod.

10.40 A uniform metal disk of mass m and radius R lays on the ice at a hockey rink. A top view of the disk is shown at right. The normal force balances the weight of the disk. Friction is negligible. In addition, three forces of equal magnitude F are exerted on the disk as shown.

a) Determine the angular acceleration of the disk. Hint: the center of mass can be considered as a pivot point.
b) Determine the translational acceleration of the disk.

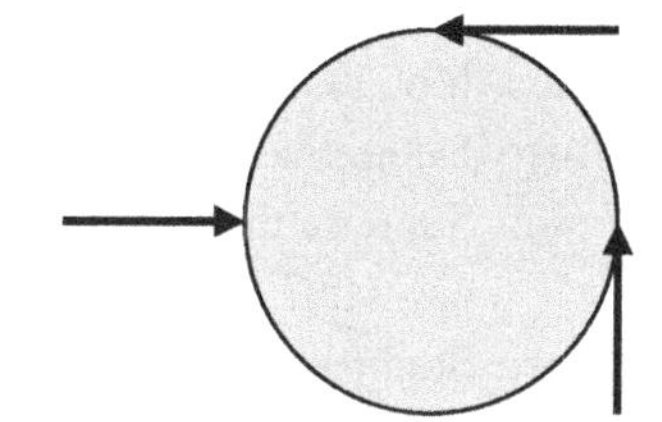

10.41 A uniform metal disk of mass m and radius R lays on the ice at a hockey rink. A top view of the disk is shown at right. The normal force balances the weight of the disk. Friction is negligible. In addition, three forces of equal magnitude F are exerted on the disk as shown.

a) Determine the angular acceleration of the disk.
b) Determine the translational acceleration of the disk.

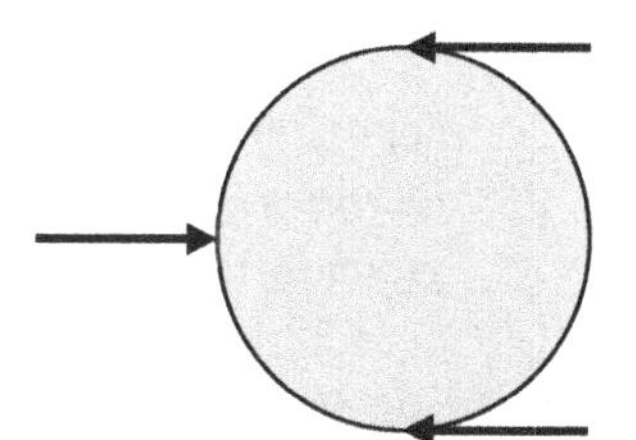

10.42 A uniform metal disk of mass m and radius R lays on the ice at a hockey rink. A top view of the disk is shown at right. The normal force balances the weight of the disk. Friction is negligible. In addition, three forces of equal magnitude F are exerted on the disk as shown.

a) Determine the angular acceleration of the disk.
b) Determine the translational acceleration of the disk.

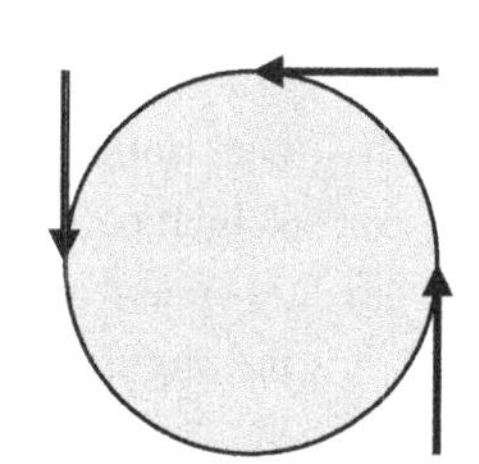

10.43 Review A spherical shell with mass 0.500 kg and radius 15.0 cm rotates about its center of mass. A plot of angular speed versus time is shown for the shell.

a) Rank the magnitude of net *torque* on the shell at 0.5 s, 2 s, 4.5 s, 5.5 s, and 7 s.
b) At what time do points on the rim of the shell experience the greatest net *force*?
c) At what time during, the time interval shown, does the shell return to its initial angular position?

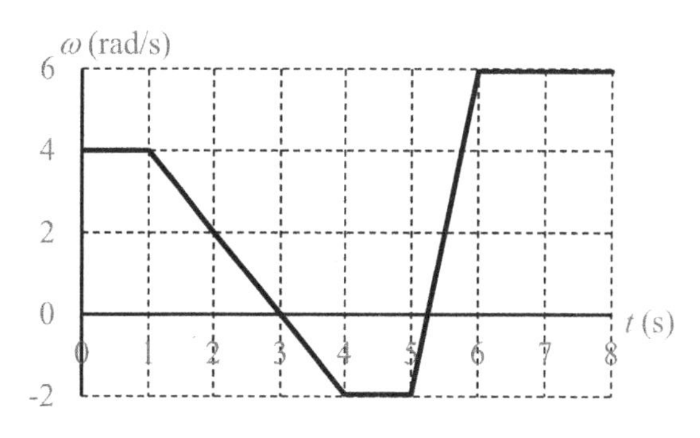

10.44 Imagine you are ripping a piece of toilet paper off the roll. If the roll is full, a quick jerk can tear off a piece of toilet paper and the roll will barely move. If the roll is almost empty, applying a quick jerk to tear off the paper may not work and the roll will simply unravel. Explain why.

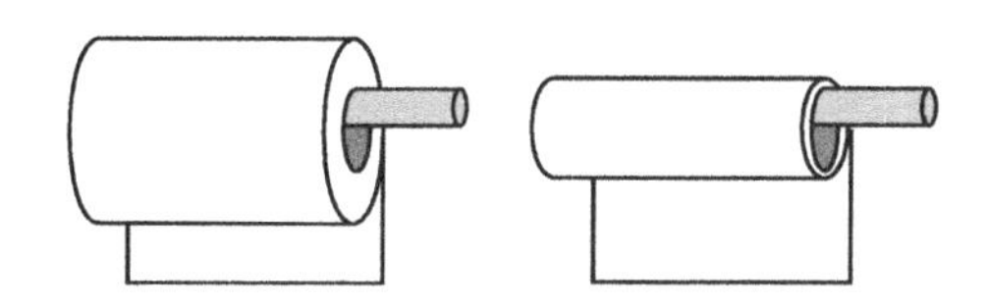

10.45 Do the demo, skip the problem…Feel the Torque A T-rod is constructed from PVC as shown in the figure. A point mass M can be connected to any one of the hooks below the rod. The left end of the rod has cylindrical handles for a person to grab the rod and try to hold it with arms parallel to the ground. The mass of the handle portion is negligible. The rod with hooks has mass m and length L. The diameter of the handles is d. Assume the rod is held parallel to the floor with an axis of rotation through the center of the handles. Assume the point mass is distance x from the axis.

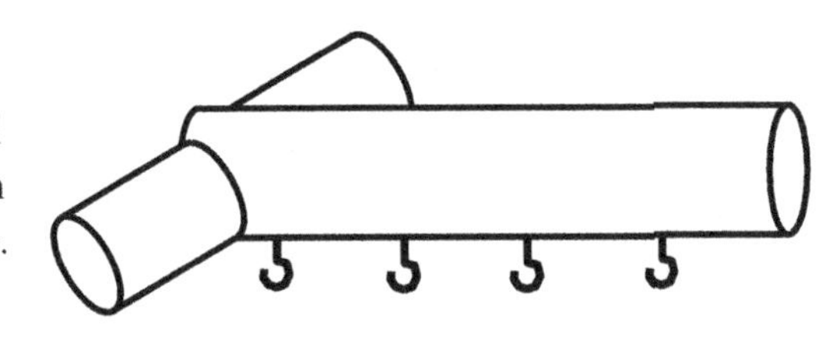

a) Determine the frictional torque required from the person's hands to keep the system in equilibrium.
b) Suppose the point mass moves farther from the axis. The weight of the system is constant. Is torque constant? Explain why or why not.
c) Suppose the rod was angled instead of parallel to the ground. The weight remains constant. Should it be easier, harder, or the same difficulty to hold the rod in equilibrium? Is it impossible to determine without more information? Don't just answer, explain why!
d) If you are interested in determining your grip strength, assume the magnitude of frictional torque is given by $\frac{\mu_s n d}{2}$. The coefficient of friction between human skin (recently washed and dried) and PVC is approximately 0.5. Typically we use $M = 1.00$ kg, $m = 0.825$ kg, $L = 1.24$ m, $d = 4.0$ cm and hook locations 8.0 cm, 38.0 cm, 68.0 cm, and 98.0 cm from the pivot. Don't forget, holding it parallel to the ground with your arms straight out is a tremendous shoulder workout as well…

10.46 Faster than g Part 1 A meterstick (mass m and length L) is pivoted at one end and held parallel to the floor. A number of pennies are at rest on the meterstick. The meterstick is released from rest.

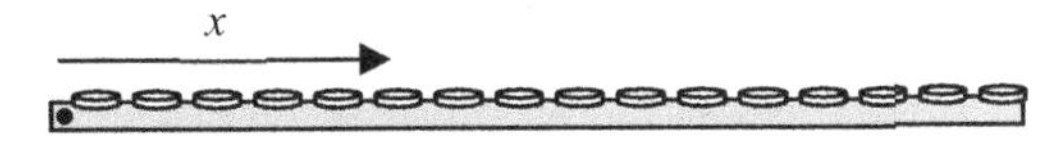

a) Ignoring the pennies, determine the *angular* acceleration of the meterstick at the instant it is released.
b) Determine the *acceleration* of a point on the meterstick distance x from the pivot.
c) Which pennies will remain in contact with the rod just after release? Explain.
d) At time t after release the rod is still swinging downward but is angled from the horizontal. Do you expect the angular acceleration of the rod to increase, decrease or stay the same? Can constant acceleration kinematics equations be applied to the swinging rod?
e) If 2-meter rod was used instead of a meterstick, how, if at all, is *angular* acceleration affected?
f) Typically, if pennies are placed every few cm or so, the total mass of the pennies is comparable to the meterstick. A brief instant after release the pennies near the pivot are still in contact with the meterstick while the pennies at the other end are not. Strictly speaking how should this affect α?

10.47 Faster than g Part 2 A board has mass m and length L. The board is initially at rest at angle θin the position shown. At the end of the board a golf ball is at rest on a tee. Near the end of the board a cup is securely affixed to the board. The cup and ball have negligible mass compared to the board. **Notice the ball is initially lower than the lip of the cup.**

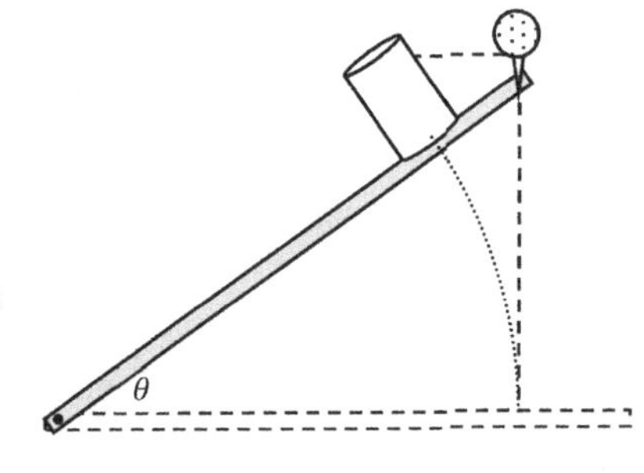

a) Our goal is to have the ball fall straight down into the cup. How far from the pivot must we place the cup? Answer in terms of L and θ.
b) What is the largest angle we can use such that the cup has an initial acceleration greater than g?
c) As the cup falls, does the acceleration of the cup tend to increase or decrease? Explain.

Using Energy with Rotation

Doing problems with energy is nearly identical to Chapter 8 with the additional of a rotational kinetic energy term on each side. Specific tips for handling different types of energy problems are discussed below.

Swinging motion

Earlier in the semester we used simple pendulums swinging. For a *simple* pendulum (point mass on end of massless string), translational motion just so happens to equal rotational motion. For anything other than a point mass on the end of a string, think of swinging motion as pure rotation. In fact, now that you know the moment of inertia of a point mass, you can view point masses that swing as pure rotation as well. For the swinging pendulum at right the energy equation would look like this:

$$m_{tot}gh_{i\text{CM}} + \frac{1}{2}I\omega_i^2 + W_{non-con} = m_{tot}gh_{f\text{CM}} + \frac{1}{2}I\omega_f^2$$

- The gravitational energy is associated with the location of the center of mass!
- For angles from the vertical we usually end up with $h_{\text{CM}} = r_{\text{CM}}(1 - \cos\theta)$.
- Typically work is zero unless axle friction is present.
- Forces towards the center of circular motion do no work.
- Translational velocity of any point on an object in circular motion is given by $v = r\omega$.

10.48 Swinging Pendulum A uniform rod of mass m and length L is supported by a pivot with negligible axle friction. The rod is initially held at angle θ from the vertical and released from rest.

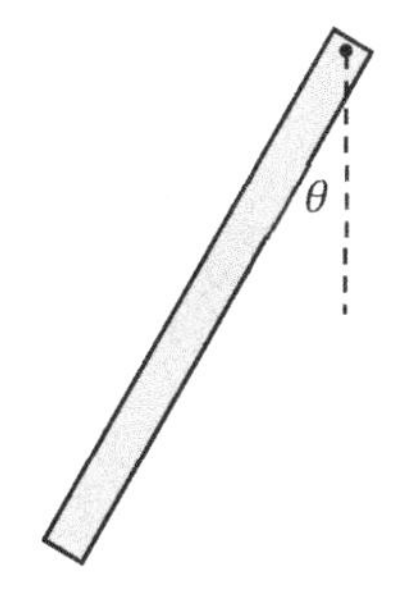

a) What is the speed of the end of the rod at the lowest point in the swing?
b) Does the answer makes sense in the limit θ goes to 0?
c) *Just after release*, which way does the acceleration of the end of the rod point?
d) At the lowest point in the swing, which way does acceleration of the end of the rod point?
e) Assume $\theta = 90°$ and $L = 1.00$ m. Determine the *translational* speed of the tip of the rod at the bottom of the swing. Think: do you expect your result to be bigger, smaller, or equal to $\sqrt{2gL}$?
f) **Going further:** what is force exerted by the pivot point on the rod as the rod goes through the lowest point in the swing? Before answering, Think: should the answer be more, less or equal to mg?

10.49 Class Demo Two rods are each equipped with a frictionless pivot at one end. The rods are each held parallel to the floor and simultaneously released from rest. The rods are made from the same material and one is longer.

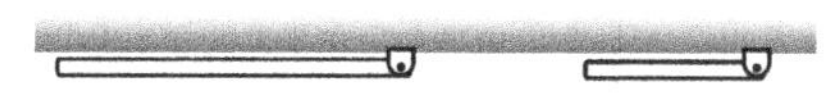

- Which rod will swing to the bottom first?
  Is the race a tie? Explain!!!!
  Take your best guess, then look at the solution.
- Support your answer by determining the speed of each rod's center of mass as it reaches the lowest point in the swing.
- Compare each speed to that of a freely falling mass that travels the same vertical distance.
  Think: which *should* be larger *and why*?

10.50 A uniform rod of mass m and length L is supported by a pivot with negligible axle friction. A point mass m is attached to rod distance x from the pivot. The object is initially held at an angle of θ from the vertical and released from rest.

a) Determine the distance from the pivot to the center of mass of the rod-point mass system.
b) What is the speed of the point mass at the lowest point in the swing?
c) **Challenge:** Assume $\theta = 90°$ and $L = 1.00$ m. Plot the rotation rate at the bottom of the swing (ω_f) versus x.
d) **Challenge:** What value of x gives maximum rotation rate at the bottom of the swing? Is it $x = 0$, $x = L$, or something in between? Discuss with your neighbor. Now determine the value. To make the equation less messy you may set $\theta = 90°$.
e) If $\theta = 90°$ and $x = L$ should the speed of the point mass be $\sqrt{2gL}$ at the bottom of the swing? Explain and verify your result from part b matches your expectation.

10.51 A uniform disk of radius R initially has mass M. A hole of radius $R/2$ cut from it as shown. The disk is then attached to an axle with negligible friction (black dot in figure, axis points out of page). The disk-with-hole is then released from rest. THINK: explain why the disk with hole rotates as shown in the figure.

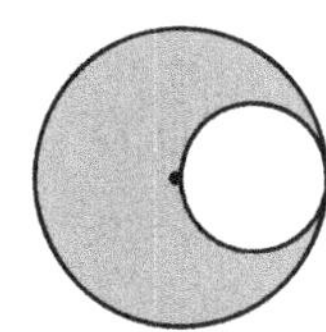

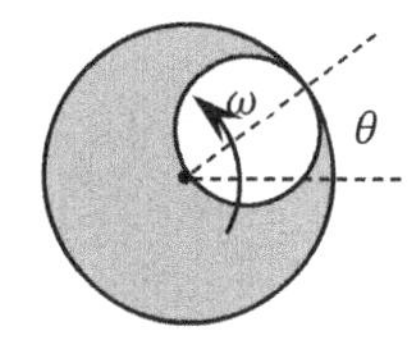

a) Determine the mass of the disk-with hole. Answer as a fraction times M
b) Determine the initial center of mass location (after hole cut but before release).
c) Determien the moment of inertia after the hole is cut Answer as a simplififed fraction times MR^2.
d) Determine the rotation rate of the disk-with-hole after it has rotated through angle θ.
e) Determine the angular acceleration of the disk-with-hole after it has rotated through angle θ.
f) **Going further:** what force is exerted by the pivot point as the center of mass goes through the lowest point in the swing? Before answering, Think: should the answer be more, less or equal to mg?

Note to teachers: in solutions for next few pages the left block is purple while the right block is red.

Pulley *Rotates* While Blocks *Translate*

Consider the figure at right where two blocks are attached at different points to a pulley. The following is useful to think about

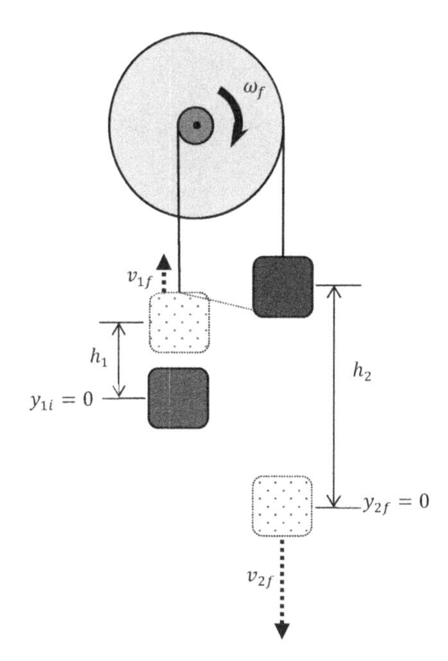

- The pulley doesn't move up and down; no need to worry about *gravitational* potential energy or *translational* kinetic energy for the pulley.
- The blocks are NOT rotating. No need to worry about *rotational* kinetic energy for the blocks.
- Typically tension will be an internal force. Tension typically does no NET work on the blocks-and-pulley system.
- Unless otherwise specified, we typically assume axle friction is negligible. If NOT negligible, axle friction would contribute to $W_{non-con}$ to the energy equation.
- The arclength of string unraveled from the pulley is typically equal to the height displaced by the blocks. This assumes the strings do not slip on the edge of the pulley which is usually valid.

Most important to understand

- When strings connect at differing radii, the acceleration, velocity, and displacements of the blocks ARE NOT EQUAL!!!! Even though the blocks connect at different radii, both blocks DO have the same ANGULAR acceleration, ANGULAR velocity, and ANGULAR displacement.

Massive Pulley Atwood's Machine Using Torque

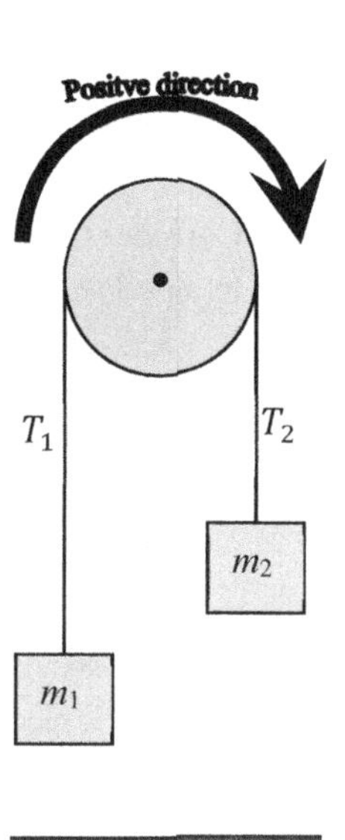

Earlier in the semester we had an Atwood's machine when the pulley was massless and frictionless. For reference we know that the acceleration for the *massless* pulley case is

$$a = g\frac{m_2 - m_1}{m_2 + m_1}$$

and the tension in the string is given by

$$T = \frac{2m_2 m_1 g}{m_2 + m_1}$$

10.52 Atwood's with torque methods Now we want to consider a pulley with non-negligible mass. Axle friction is still negligible but now the pulley of radius *R* has non-zero moment of inertia *I* about the axis shown. Because the pulley has mass, the tension differs from one side of the pulley to the other as shown. The string is still massless and inextensible which does allow us to say the blocks accelerate with equal magnitude. Lastly, the string does not slip on the pulley. This allows us to relate the blocks' translational acceleration to the pulley's angular acceleration using $a = R\alpha$.

a) Show the acceleration with a *massive* pulley is

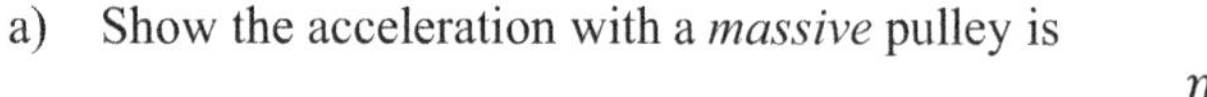

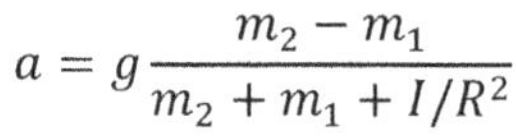

$$a = g\frac{m_2 - m_1}{m_2 + m_1 + I/R^2}$$

b) Verify the result matches our previous result if the mass of the pulley is set to zero.

c) Determine each tension. Verify the results match the massless pulley derivation when $I = 0$.

10.53 Atwood's with energy methods Reconsider the previous problem using energy and kinematics.

a) What assumptions do we make in order to assume the acceleration of the system is constant?

b) Assume the system is released from rest. Use energy methods to determine the speed of m_2 after it has fallen distance h.

c) Use kinematics to determine the acceleration of m_2 after it is released from rest.

d) Think: if there was no friction on the edge of the pulley (where string makes contact) the string would just slide right over it and the pulley wouldn't rotate at all. At the same time, the derivation you just completed assumed there was no external work done. What the heck is going on here? Is this answer just an approximation or do you think it is ok to assume the work done by friction at the edge of the pulley is zero?

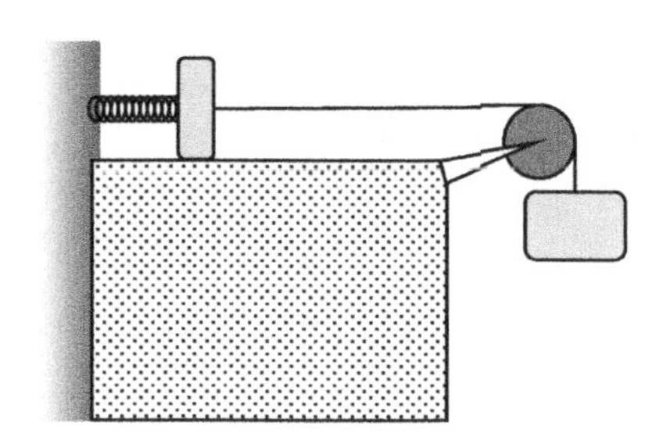

10.54 A block of mass m is attached to a wall with a light spring of constant k. The spring is initially unstretched. The block is attached to a hanging block with three times the mass using a light, inextensible string over a pulley with negligible axle friction. The pulley is essentially a uniform disk. Upon being released from rest, the block on the level surface slides with negligible friction. When the system first reaches the equilibrium position, it is still moving and overshoots the equilibrium point! When the system is equilibrium the *acceleration* of the system is zero but the speed of the hanging block is v! Determine the mass of the pulley.

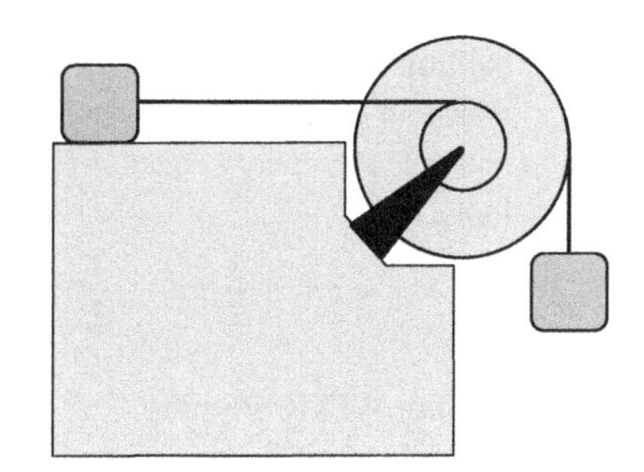

10.55 Two blocks of identical mass m are connected to the pulley using light, inextensible springs and the system is released from rest. The pulley is made by welding a two thin disks together. The larger disk has radius R while the smaller disk has radius $0.40R$. Once combined, the total moment of inertia of the pulley about its axis is $I = 0.725mR^2$. Axle friction is negligible. There is negligible friction between the left block and the horizontal surface. Assume neither string slips on the pulley.
Determine the translational acceleration of the hanging mass after the system is released. **Answer as a number with 3 sig figs times g.**

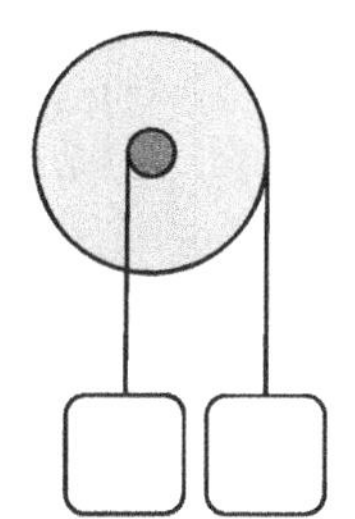

10.56 Two identical blocks of mass m are connected to a pulley of outer radius R using light, inextensible strings. The pulley also has mass m with moment of inertia $I = kmR^2$. Note: k is some constant with a value between 0 and 1. The right block is connected to the outer radius R of the pulley while the left block is connected at radius $R/3$. The angular acceleration of the pulley is observed to be α. Figure not to scale.

a) Which way, if any will the pulley turn?
b) Determine the ratio of the acceleration of the left block to right block? Note: if they have the same acceleration, the ratio is 1. If they do not accelerate, answer zero.
c) Are the tensions of each cable equal to mg? Explain.
d) Will the blocks travel the same distance? Will they rotate through the same angle? Explain.
e) Determine the constant k in terms of the givens. Show $k = \frac{2g}{3R\alpha} - \frac{10}{9}$.

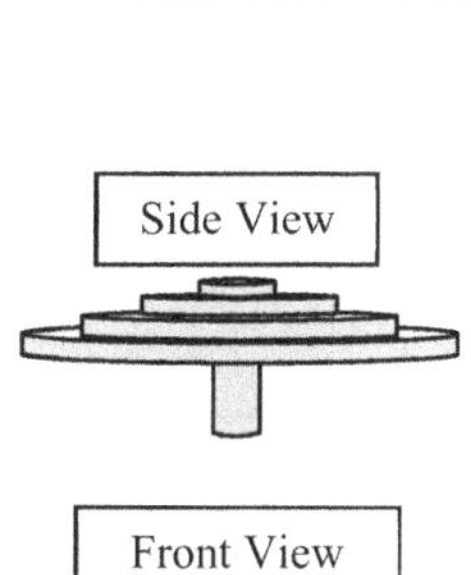

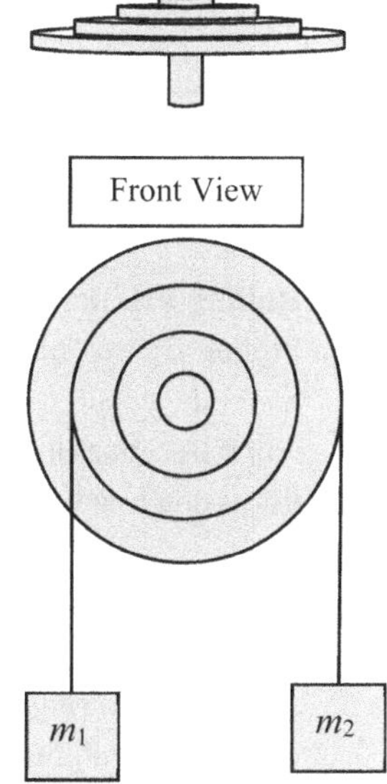

10.57 Using torque methods Our massive pulley actually looks like the figures shown at right. The mass of the pulley is 176 g. Strings can be attached at 10 mm, 20 mm, 40 mm, or 60 mm. The strings attaching the masses to the pulley are wound around the pulley several times to ensure no slipping occurs. Suppose m_2 is connected to r_2 while m_1 is connected to r_1. The system is released from rest.

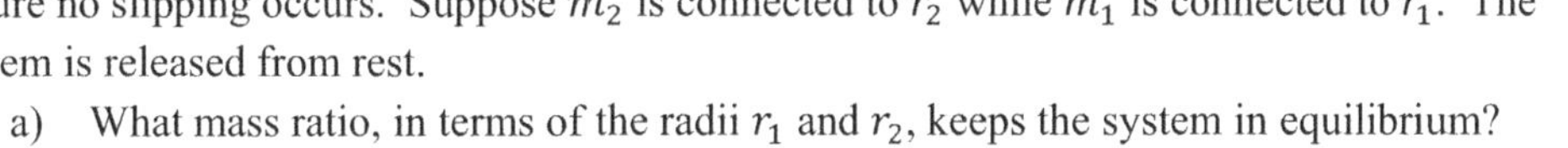

a) What mass ratio, in terms of the radii r_1 and r_2, keeps the system in equilibrium?
b) Suppose m_2 falls distance h_2 in time t. How far does m_1 travel upwards?
c) Relate the acceleration magnitudes of the two masses.
d) Determine the moment of inertia of the pulley in terms of m_1, m_2, r_1, r_2, a_2, and g.
e) Using $m_2 = 200$ g, $m_1 = 150$ g, $r_2 = 60$ mm, and $r_1 = 40$ mm I found

$$I_{exp} = 7.2 \times 10^{-4}\ \text{kg} \cdot \text{m}^2 \left\{\frac{g}{2a_2} - 1.333\right\}$$

Use a stopwatch and a meterstick in lecture to measure h_2 and t. From kinematics we know $a_2 = \frac{2h_2}{t^2}$. Determine I_{exp}. If possible, try to time a fall longer than 1.0 m.
For comparison, if the pulley is a solid disk we find $I = \frac{1}{2}MR^2 \approx 3.17 \times 10^{-4}\ \text{kg} \cdot \text{m}^2$.

10.58 Using energy methods Reconsider the previous problem using energy. Assume the system is released from rest. Determine the *rotational* speed of the pulley after m_2 has fallen distance h_2. **WATCH OUT!** In this problem the masses will not move the same amount, they will not have the same speed, and they do not have the same accelerations. Notice, however, the strings are connected to the same axis and do experience the same *angular* displacement, same *angular* velocity, and same *angular* acceleration.

Rolling Motion

When an object rolls it has both rotational energy *and* translational energy. The energy equation looks for a ball rolling down an incline looks like

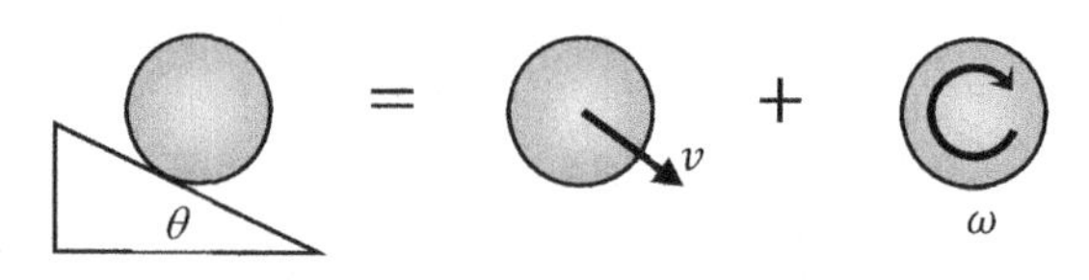

$$mgh_i + \frac{1}{2}mv_i^2 + \frac{1}{2}I\omega_i^2 + Work = mgh_f + \frac{1}{2}mv_f^2 + \frac{1}{2}I\omega_f^2$$

Rolling WITH Slipping

For rolling with slipping, one must consider the center of the rolling object as the axis of rotation. Usually one is required to do force, torques, and both translational and rotational kinematics. An example is given later.

Rolling WITHOUT Slipping

This is the more common of the two rolling motions. Unless otherwise specified, assume all rolling motion is rolling *without* slipping. Energy problems are typically useful in rolling without slipping. When rolling without slipping occurs the following can be used

1) $v = r\omega$
2) $a = r\alpha$
3) $\Delta x = s = r\Delta\theta$
4) Static friction is present but usually $f < \mu_s n$. **Do not** expect to use $f = \mu_s n$.
5) Static friction does no work (*if no slipping*)
6) Center of mass can be used as the axis of rotation for torque problems.
7) The point of contact with the road is not moving! As far as translations acceleration of the center of mass is concerned, the problem is actually mathematically identical to the object swinging about the point of contact with the road. This technique/math trick is called using the *instantaneous pivot*. Note: since the contact point is not at the center of mass one must use the parallel axis them to determine I. This point is convenient since friction causes no torque about the instantaneous pivot. This technique fails to show the acceleration of the point in contact with the road is actually vertically upwards! HEY! This is a math trick that gets the job done quickly. If you want to be a master of this…take a dynamics class will ya?!?!?!

DO PROBLEM 10.44 and thoroughly master all subtleties of this solution including the extra notes after the solution.

Rolling Friction vs Sliding Friction During Rolling Motion

Sliding friction must be present else rolling motion would not occur. Without it, there is no force to cause a torque about the center of mass and cause rotation! **However, when rolling *without* slipping, sliding friction does no work because the point of contact is not in relative motion with the ground.** Rolling objects eventually slow down due to work done by *rolling* friction. Unless otherwise specified, ignore work done by rolling friction.

Yo-Yo Problems

A yo-yo falling down as a string unravels is essentially the same kind of problem as rolling without slipping. This assumes the string unravels without slipping on the axle of the yo-yo. Friction is what keeps the string from slipping on the axle but, because the string is not displaced relative to the axle, this frictional force does no work!

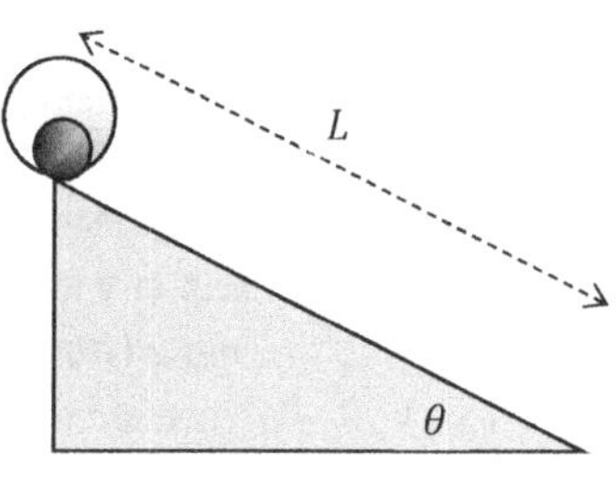

10.59 Rolling Race Objects are placed side by side on a ramp and released from rest as shown in the figure. Consider the following rolling races: 2 spheres of different sizes, 2 thin rings of different sizes, a sphere and a cylinder of equal radius, a sphere and a cylinder of equal mass, a large solid cylinder and a small thin ring.

a) Go with your gut. What factors should affect the race? Mass? Radius? Both? Neither? Shape?
b) Now do a conservation of energy problem for an object with moment of inertia I (for axis through center of mass) and radius R.
Ignore any work done by rolling friction. If the object rolls without slipping we expect $v = r\omega$. Determine the final speed in terms of m, g, L, θ, I, and R.
c) Evaluate your answer to part a by plugging in the moments of inertia for a sphere, a disk, á spherical shell, and a thin ring. What drops out? What determines which object(s) win the race?
d) Determine the *angular* acceleration of an object rolling (w/o slipping) on the incline in terms of m, g, θ, I, and R.

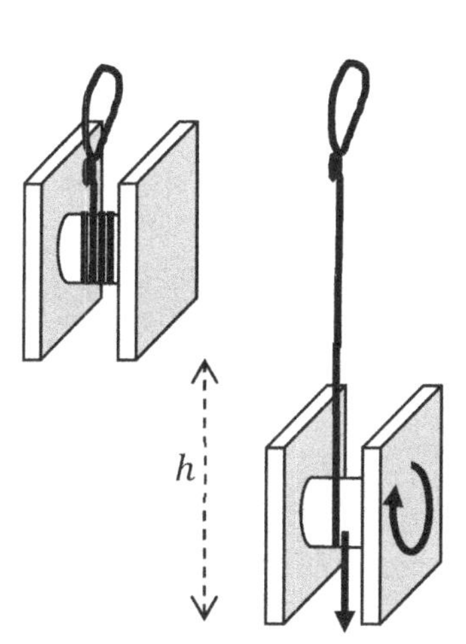

10.60 Yo-Yo A string is wrapped around a yo-yo with moment of inertia I and mass M. The center spindle of the yo-yo (where the string connects) has radius R. The yo-yo is initially held at rest. Upon being released the string wrapped around the spindle unravels without slipping.

a) As the object falls, is it in freefall? As such would you expect the acceleration to be equal to g, greater than g, or less than g?
b) Determine the rotational speed of the yo-yo after it has fallen a distance h. Answer in terms of h, I, M, g, and R.
c) Determine the translational acceleration of the cylinder after it has been released from rest. Answer in terms of I, M, g, and R.
d) Our yo-yo is comprised of two squares and a central cylinder as shown in the figure. The squares each have diagonal D = 26.7 cm and mass m_s = 421 g. The central cylinder has m_c = 55 g and diameter d = 2.86 cm. Determine the moment of inertia (about the center of mass) of the yo-yo.
e) Determine the time required for the yo-yo to fall 1.10 m. This should be the entire length of string assuming the knot/loop is just beyond the edge of the square. Do the experiment and see if it matches.
f) Is it reasonable to ignore the central cylinder in your calculations? Suppose you ignored the central cylinder. What % error does this introduce into the moment of inertia?

10.61 Rolling Spool Demo Consider the three possible ways to pull on a spool shown below. The spool is initially at rest. The straight line is a string attached to the spool. Which way, if any, will the spool roll/rotate for each case? Hint: assume it rolls without slipping. If it rolls without slipping, you could use the center of mass as the pivot or you could use…

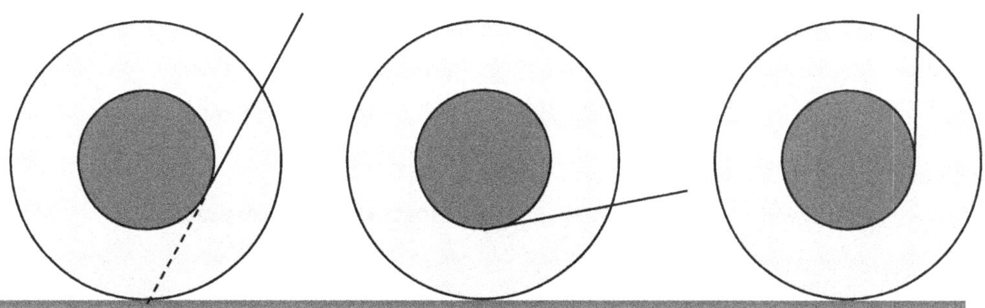

10.62 A yo-yo has center radius r and outer radius R as shown in the figure.

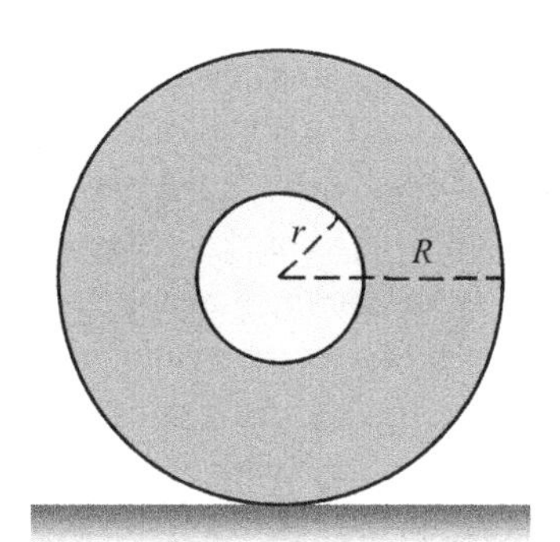

Sketch the direction one must pull on the string if one wanted to cause the yo-yo to slide without any rotation.
Determine the angle (from the vertical) at which the string is pulled and the yo-yo will *slide* instead of *roll.*

10.63 A solid sphere of negligible size ($r \ll R$) rolls without slipping on the track shown in the figure. The sphere is given a very slight push to get it moving. Assume the initial velocity is close enough to zero to be ignored.

a) Determine the sphere's speed at point **A** in terms of R, g, and h.
b) Determine the minimum height h that allows the sphere to roll through the vertical loop without losing contact with the road.
c) Compare this situation to one where the block slides with negligible friction. Which situation should require a larger initial height? Explain.

Note: if the sphere has non-negligible size, one must correctly reference the center of mass height at each position. This is annoying but not much different in terms of algebra…

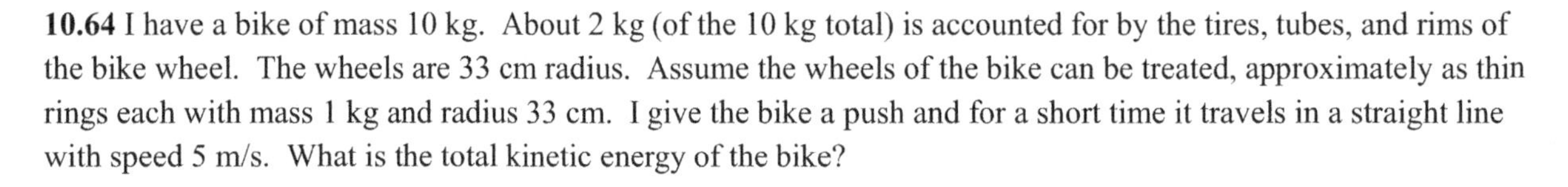

10.64 I have a bike of mass 10 kg. About 2 kg (of the 10 kg total) is accounted for by the tires, tubes, and rims of the bike wheel. The wheels are 33 cm radius. Assume the wheels of the bike can be treated, approximately as thin rings each with mass 1 kg and radius 33 cm. I give the bike a push and for a short time it travels in a straight line with speed 5 m/s. What is the total kinetic energy of the bike?

10.65 A wheel with center of mass moment of inertia $I = kmr^2$ rolls without slipping with speed v.

a) What is the speed of the point in contact with the road? Hint: consider the combined effects of translation and rotation.
b) What is the speed of the top of the wheel?
c) What fraction of the total kinetic energy is rotational?

10.66 A solid sphere starts from rest and rolls down a slight incline without slipping. The solid sphere takes time t_1 to roll down the incline. The student then asks Angus to drill a big hole through the center of the sphere. Angus drills the hole for her because she gives him a free energy drink and they can both complain about Jorstad while the work is getting done. The student then aligns the sphere-with-hole on the same incline and releases it from rest. This time the sphere takes time t_2.

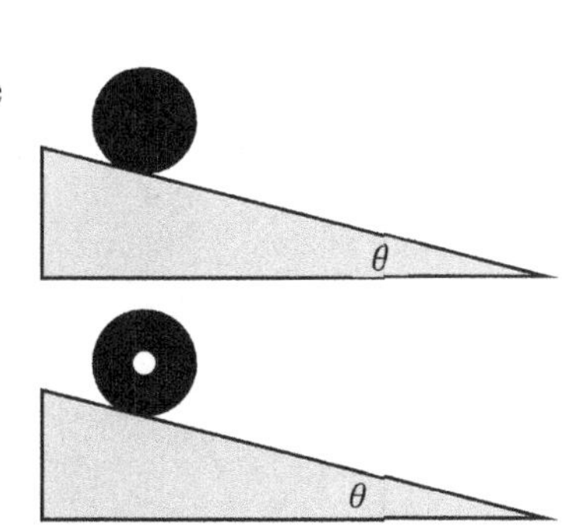

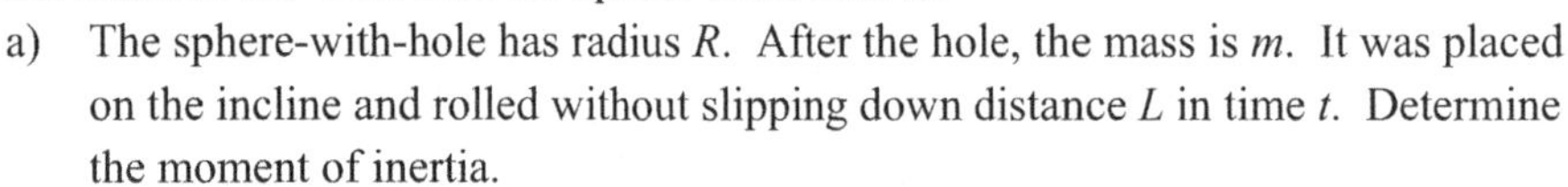

a) The sphere-with-hole has radius R. After the hole, the mass is m. It was placed on the incline and rolled without slipping down distance L in time t. Determine the moment of inertia.

b) Which time is larger? Clearly explain your reasoning.

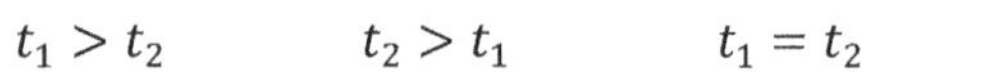

$t_1 > t_2$ $\qquad$ $t_2 > t_1$ $\qquad$ $t_1 = t_2$ $\qquad$ Impossible to determine without more info

10.67 Concept Q/Blow your mind Consider two identical disks of mass M and radius R initially at rest on a frictionless surface. You could think of the figure as being a view from above of two disks on a perfectly smooth hockey rink. A string is tied to the center of disk 1. Upon being pulled with constant force magnitude F, disk 1 accelerates with rate a. Disk 2 has the string wound around the edge of the disk. As you pull on the string, we expect disk two to move to the right but it will also start to spin.

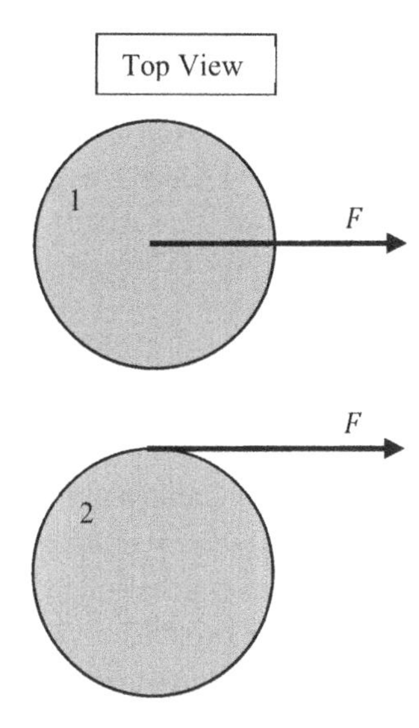

If we pull on disk 2 with the same constant force magnitude F, will it accelerate faster, slower, or at the same rate as disk 1?

Rolling WITH slipping:

1) **Until road speed matches rotation speed $v \neq r\omega$ and $a \neq r\alpha$!!!!**
2) **Can no longer use the instantaneous pivot point! Must use center of mass pivot.**

DEMO: One can demonstration of rolling *with* slipping using a hula hoop. Pretty big, easy to see in back of class.

10.68 Rolling with slipping. Assume bowling ball is a uniform sphere. Initially, the ball slides down alley with no rotation at speed v_0. The ball rolls *with* slipping until road speed matches rotation speed $(v_f = R\omega_f)$. The final speed is not known. The coefficients of friction between the ball and the lane are μ_s and μ_k.

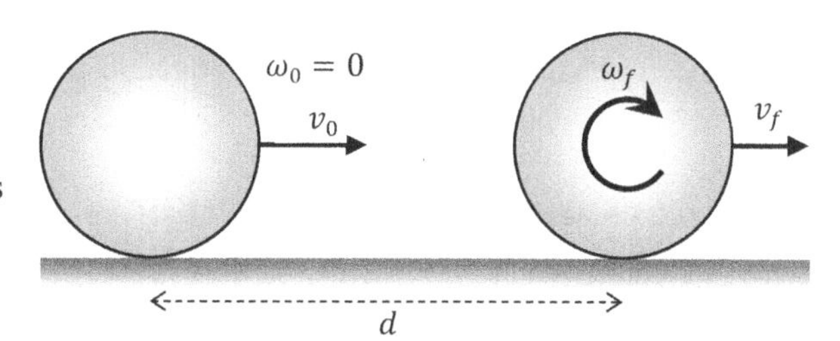

a) Determine the time the ball rolls with slipping.
b) Determine the translational speed of the ball at the moment rolling without slipping begins.
c) Determine the distance traveled by the ball.
d) Determine the total number of revolutions the ball makes during the rolling with slipping.
e) Assume $\mu_s = \mu_k = 0.12$, $m = 5.0$ kg, $r = 11$ cm, and $v_0 = 7.2\ \frac{\text{m}}{\text{s}}$. The length of a bowling lane is about 18 m. Sketch plots of v and $R\omega$ vs t on the same graph for the entire 18 m distance.

Note: this is a simplified model as bowling lanes often have unusual oiling patterns to cause different coefficients of friction on different parts of the lane. Bowling balls have off-axis center of mass causing the ball to move in a more complicated fashion. Still, this is a decent first approximation.

10.69 Challenge: A spool of wire is a thick ring with outer radius R, inner radius $\frac{R}{2}$, and mass m. The wire is pulled with a constant force of magnitue F parallel to the level surface. Assume the spool rolls without slipping. Assume the wire unravels without slipping. Assume as the wire is pulled a negligible amount of mass is pulled away from the spool.

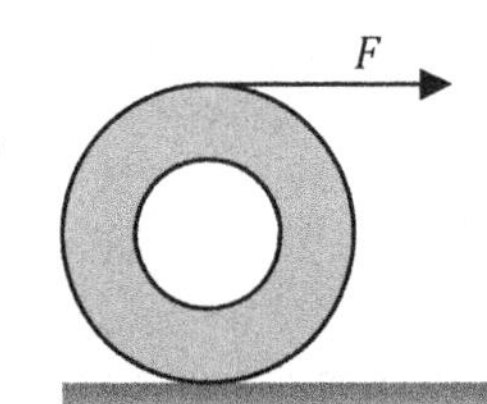

a) Determine the acceleration of the spool's center of mass.
b) Determine the magnitude and direction of the fricitonal force on the spool.
c) Determine the translational speed of the spool after the center of mass has traveled distance L.
d) Design/describe an experimental appratus that would allow you to apply a constant force to the spool.

10.70 Challenge: A solid sphere of mass m and radius r rolls without slipping in a half-pipe. The half-pipe has radius R. Notice there two angles in this problem, one is the angle θ from vertical; the other is the angle ϕ of ball as it spins. Do not assume $r \ll R$.

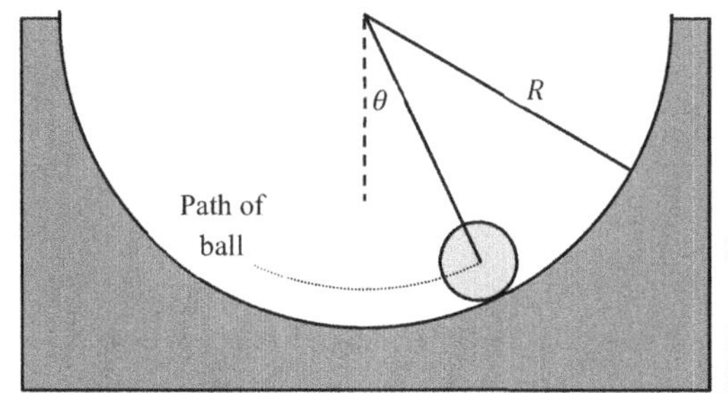

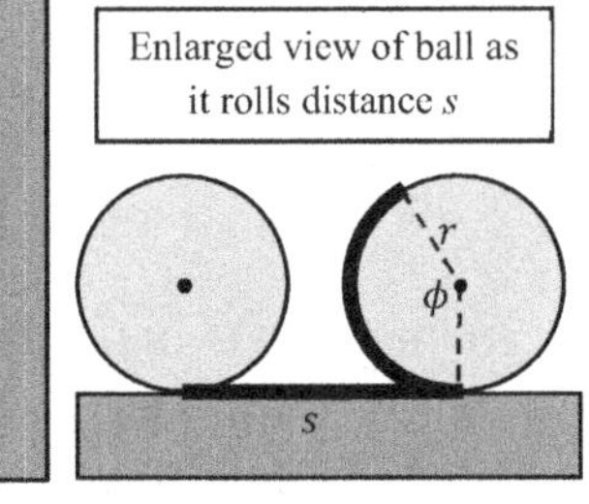

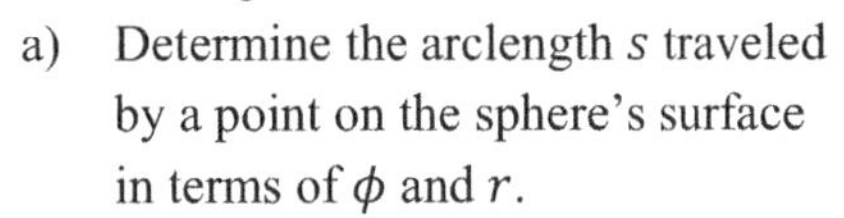

a) Determine the arclength s traveled by a point on the sphere's surface in terms of ϕ and r.
b) Determine the arclength s traveled by a point on the sphere's surface in terms of θ and R.
c) Use the above two results to define a relationship between the two angles θ and ϕ.
d) Determine the radius of circular motion traveled by the sphere's center of mass.
e) The ball is released from rest with initial angle $\theta = 60°$. Determine the translational speed of the ball at the lowest point in the half pipe.

10.71 Challenge: A plank of mass M is placed on two cylinders, each with mass m and radius R. Initially the two cylinders are equally spaced and each supports half of the weight of the plank. The plank, starting from rest, is pulled by a force with constant magnitude F parallel to the level surface. Assume that no slipping occurs between the cylinders and the plank. Assume no slipping occurs between the cylinders and the floor.

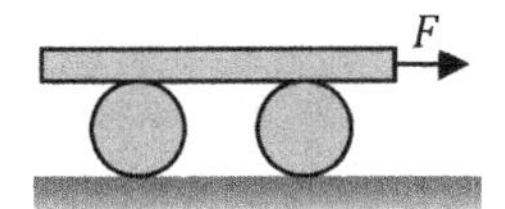

a) Determine the initial acceleration of the plank.
b) Determine the initial acceleration of the cylinders.
c) Determine the initial magnitude and direction of all frictional forces acting on the front cylinder.

10.72 Challenge: A spool of mass m has moment of inertia $I = kmR^2$ where k is a known constant. A string is attached to the inner radius R. The outer radius of the spool is $2R$. Length L of string is pulled with constant force magnitude F. Assume the spool starts from rest and rolls without slipping. Assume the string unravels without slipping as well.

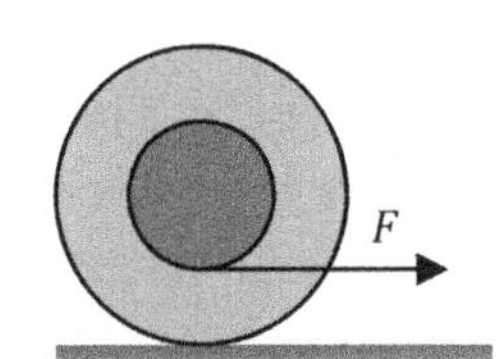

a) Determine the center of mass acceleration.
b) Determine the translational speed of the spool after length L of string has been pulled.
c) Determine magnitude and direction of the frictional force.

ANGULAR MOMENTUM

Angular momentum is conserved for systems with no external *torque*. In general we expect $\Sigma\vec{\tau}_{ext} = \frac{d\vec{L}}{dt}$.
For comparison, remember *linear* momentum was conserved in case of no external *force*. If external force was non-zero (and mass was constant) we used $\Sigma\vec{F}_{ext} = \frac{d\vec{p}}{dt} = \frac{d(m\vec{v})}{dt} = m\frac{d\vec{v}}{dt} = m\vec{a}$.

For a point particle use

$$\vec{L} = \vec{r} \times \vec{p} = \vec{r} \times m\vec{v} = mvr \sin\phi \ (direction\ from\ RHR)$$

- $\vec{r}$ is the vector from an arbitrarily chosen point (usually a pivot point) to the moving point mass
- ϕ is the angle between the direction of $\vec{v}$ & $\vec{r}$. See the figure below to make sense of all these words
- direction of angular momentum will be given by a right hand rule (just like torque).

Magnitude Trick 1: $L = mvr_{\perp}$

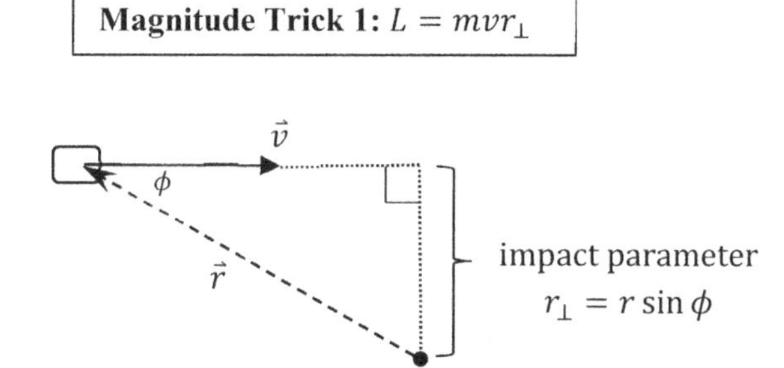

Magnitude Trick 2: $L = mv_{\perp}r$

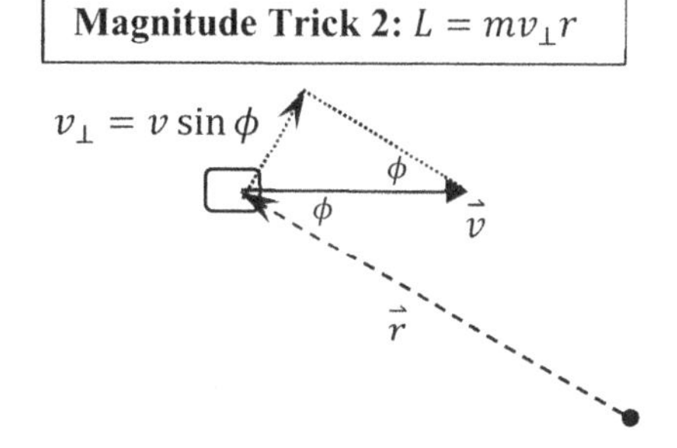

For extended objects (things other than point particles) we typically use $L = I\omega$.

Special note regarding point masses: use $L_{point\ mass} = mvr_{\perp}$ for point masses moving in a straight line.
Use $L_{point\ mass} = I\omega = (mr^2)\omega$ for a point mass attached to a rotating object distance r from the pivot.
Note: for point particle in circular motion (with center of the circle as the pivot) we find $L = mvr = mr^2\omega$

Important Relationship

Torque causes a change in rotation. Changes in rotation imply angular momentum is changing.

$$\frac{\Delta\vec{L}}{\Delta t} = \frac{\vec{L}_f - \vec{L}_i}{\Delta t} = \Sigma\vec{\tau}_{external}$$

Conservation of angular momentum may be used for problems with no net external torque.

If external torque is zero (or negligible) the above result simplifies dramatically to

$$\vec{L}_f = \vec{L}_i$$

In practice this equation takes one of two forms shown below. Note: usually a $\hat{k}$ is implied on all terms.

One or two objects rotating with collision	Rotating object changes shape
$L_{1i} + L_{2i} = L_{1f} + L_{2f}$ $I_1\omega_{1i} + I_2\omega_{2i} = I_1\omega_{1f} + I_2\omega_{2f}$	$L_i = L_f$ $I_i\omega_i = I_f\omega_f$

Remember when we talked about coding using momentum? Here is the equivalent using angular momentum:

- Calculate net *torque* vector based on current *angular* position
- Update the *angular* momentum of the system using $\vec{L}_f = \vec{L}_i + \vec{\tau}_{net}\Delta t$
- Compute current average *angular* velocity using $\vec{\omega}_{avg} \approx \frac{\vec{L}_f}{I}$
- Update the angular position using $\vec{\theta}_f = \vec{\theta}_i + \vec{\omega}_{avg}\Delta t$
- **Strictly speaking we may only treat $\boldsymbol{\theta}$ as a vector for rotations in a single plane (about one axis).**

11.1 Consider a moon in a circular orbit of radius R about a planet. The moon has mass m and speed v. The moon is the small grey circle on the right side of the orbit. No mass is present at point **A**; it is labeled for ease of communication. The planet is centered on the point labeled **B**. As the moon orbits the planet, you may assume the mass of the moon is very small compared to the planet. The center of mass of the moon-planet system is essentially at point **B**. The only force acting on the moon is a force of gravitational attraction towards the planet. Assume the orbit lies in the standard xy-plane (and thus the positive z-axis points out of the page).

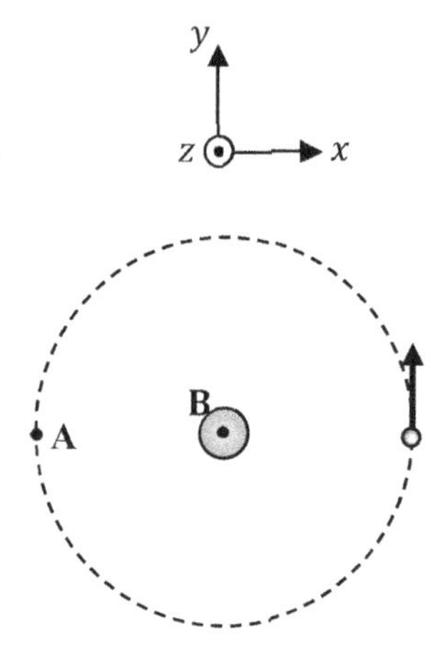

a) For the instant shown, determine the angular momentum of the moon about point **A**.
b) For the instant shown, determine the angular momentum of the moon about the center of the orbit (about point **B**).
c) For the same instant (moon at top of circle), determine the angular momentum of the moon about point **A**.
d) Now consider the instant in time when the moon reaches the top of the circle. Determine the angular momentum of the moon about point **B**.
e) We can choose any pivot we want. Which seems more convenient to work with to you/why care?

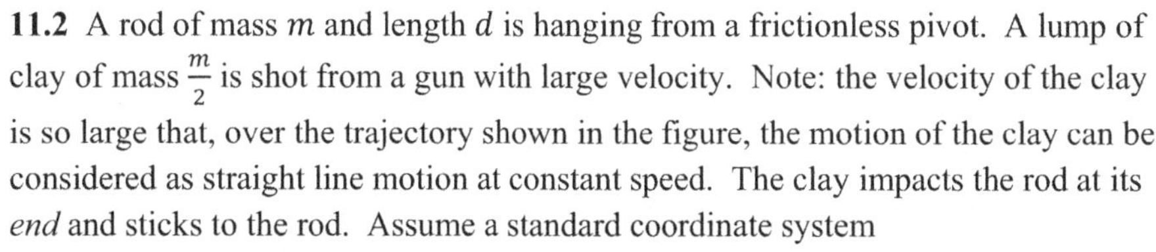

11.2 A rod of mass m and length d is hanging from a frictionless pivot. A lump of clay of mass $\frac{m}{2}$ is shot from a gun with large velocity. Note: the velocity of the clay is so large that, over the trajectory shown in the figure, the motion of the clay can be considered as straight line motion at constant speed. The clay impacts the rod at its *end* and sticks to the rod. Assume a standard coordinate system

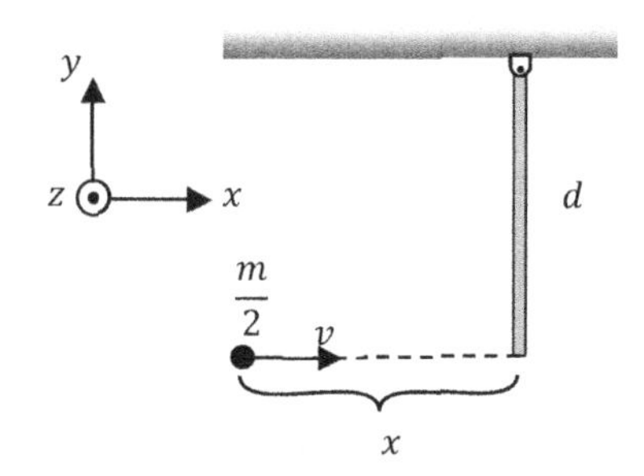

a) What is the best choice of arbitrary point we should use in determining angular momentum?
b) Determine the impact parameter about this point.
c) What is the initial angular momentum of the clay about the pivot point?

11.3 A rod of mass m and length d is hanging from a frictionless pivot. A lump of clay of mass $\frac{m}{2}$ is shot from a gun with an <u>unknown</u> large speed v. Note: v is so large that, over the trajectory shown in the figure, the motion of the clay can be considered as straight line motion at constant speed. The clay impacts the rod at its *center* and sticks to the rod. After the collision the rod swings up to a maximum angle of 90° (parallel to the ceiling).

a) Assuming the real life pivot is our pivot point for angular momentum, determine the impact parameter of the clay. If this is an in class activity and you finish this part early, continue with parts b) through g) and see how far you can get. Otherwise, the rest of this problem is better to do after you make it through **11.8**.
b) Determine the initial angular momentum of the clay. Answer in terms of m, d & the unknown v.
c) What is the moment of inertia of the rod + clay system after the collision?
d) What is the rotation rate of the rod + clay system *just after* the collision (in terms of v & d) ?
e) What assumptions were made to use conservation of angular momentum in the previous step?
f) How much does the center of mass of the rod + clay system rise after the collision (when it reaches max height)?
g) Determine the initial speed of the clay just before impact with the rod.

Important Relationship

Torque causes a change in rotation. Changes in rotation imply angular momentum is changing. These concepts are related by the equation

$$\frac{\Delta\vec{L}}{\Delta t} = \frac{\vec{L}_f - \vec{L}_i}{\Delta t} = \Sigma\vec{\tau}_{external}$$

11.4 Good exercise for home? Could you code this up as a simulation? Assume the system is initially at rest (left picture). The pulley is essentially a uniform disk of radius R. In the right picture both blocks are moving with the same final speed v and the pulley rotates with angular speed ω. The initial angular momentum of system (the two blocks and pulley) is zero. Assume axle friction is negligible.

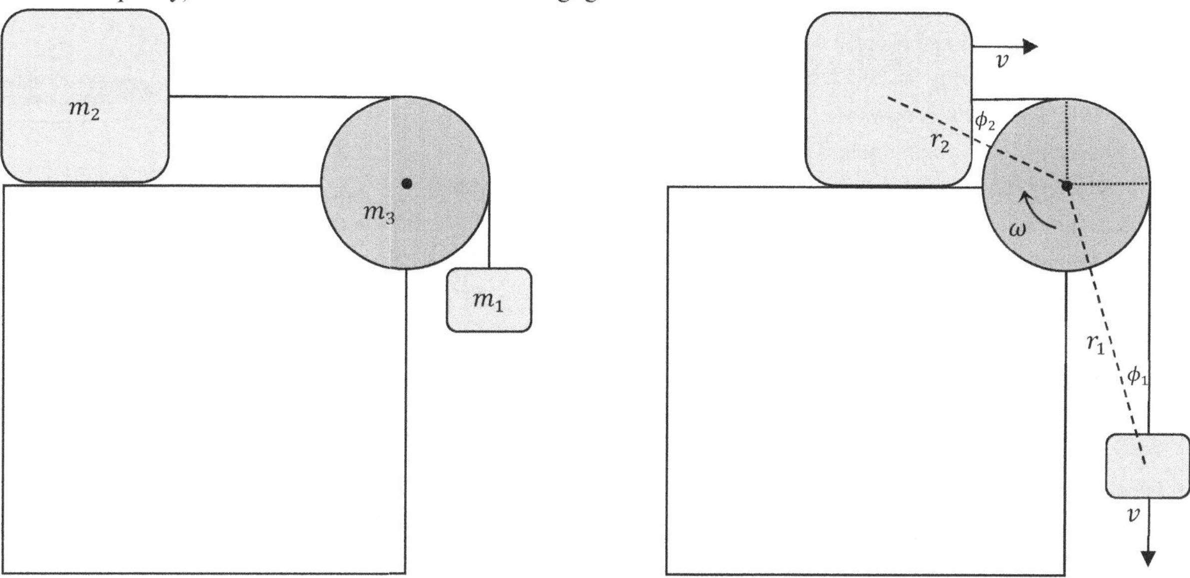

a) Use FBDs (on m_1 and m_2) and torques (on m_3) to show $\alpha = \frac{m_1 g}{R\left(m_1+m_2+\frac{1}{2}m_3\right)}$.

Remember, $T_1 \neq T_2$ because the pulley has mass!

b) Show that after some small time t the final speed of each block is given by $v = \frac{m_1 g t}{m_1+m_2+\frac{1}{2}m_3}$.

c) Show that the final angular momentum of the system is given by $L_f = \left(m_1 + m_2 + \frac{1}{2}m_3\right)vR$. Hint: show the impact parameter for each block is $r\sin\phi = R$.

d) Compare the initial and final angular momentum in the two blocks & pulley system. Is angular momentum conserved (does $L_i = L_f$)?

e) Show that $\frac{\Delta L}{\Delta t} = \frac{L_f - L_i}{\Delta t} = \frac{L_f}{t} = the\ sum\ of\ external\ torques$! Hint: in this problem the only unbalanced external torque is caused by gravity acting on the hanging block.

f) Try to code this up. For the blocks use force to update momentum & momentum to update position. For the disk, draw a line through an arbitrary diameter then code the rotation of that line using the suggested method listed below.

Remember when we talked about coding using momentum? Here is the equivalent using angular momentum:

- Calculate net *torque* vector based on current *angular* position
- Update the *angular* momentum of the system using $\vec{L}_f = \vec{L}_i + \vec{\tau}_{net}\Delta t$
- Compute current average *angular* velocity using $\vec{\omega}_{avg} \approx \frac{\vec{L}_f}{I}$
- Update the angular position using $\vec{\theta}_f = \vec{\theta}_i + \vec{\omega}_{avg}\Delta t$
- **Strictly speaking we may only treat $\boldsymbol{\theta}$ as a vector when all rotations occur about the same axis (rotation in a single plane).**

Conservation of angular momentum may be used for problems with no net external torque.
In practice this equation takes one of two forms:

One or two objects rotating with collision	Rotating object changes shape
$L_{1i} + L_{2i} = L_{1f} + L_{2f}$ $I_1\omega_{1i} + I_2\omega_{2i} = I_1\omega_{1f} + I_2\omega_{2f}$	$L_i = L_f$ $I_i\omega_i = I_f\omega_f$

Special note regarding point masses: you can choose $L_{point\ mass} = mvr_\perp$ for point masses moving in a straight line or $L_{point\ mass} = I\omega = (mr^2)\omega$ for a point mass attached to a rotating object.

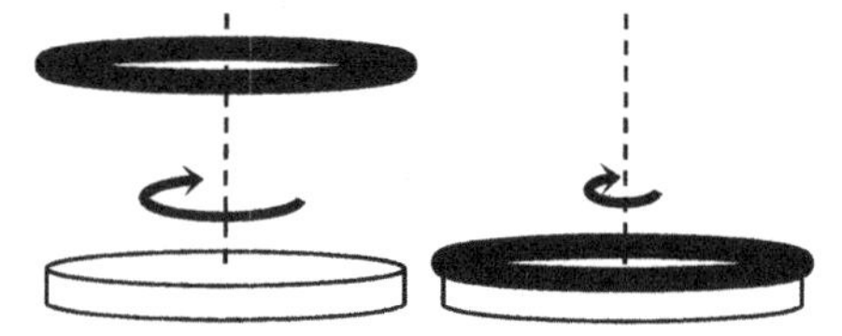

11.5 A turntable rotates with rate 10.0 RPM. The turntable can be modeled as a uniform disk of mass M and radius R. A physics student takes a thin ring with identical mass and radius and drops it on top of the turning disk. After a very brief period of time, the ring and the turntable reach equilibrium and rotate with same final rate. Assume $\hat{k}$ is upwards.

a) Determine the initial angular velocity ω_0 in rad/s.
b) Before going on, take a guess at the final rotation rate of the system.
c) Calculate the final rotation rate of the system. Answer as a fraction times ω_0 <u>then</u> compute the number.
d) Do you expect the energy of the system to increase, decrease, or stay the same?
e) Now determine the percent change in energy. To do this use the formula $\%\Delta E = \frac{E_f - E_i}{E_i} \times 100\%$. Note: an answer of zero implies the energy stayed the same, a positive answer implies energy increased, and a negative answer implies energy decreased. Compare to your intuition for the previous part…the answer often surprises people! Think: what did work to change the energy?

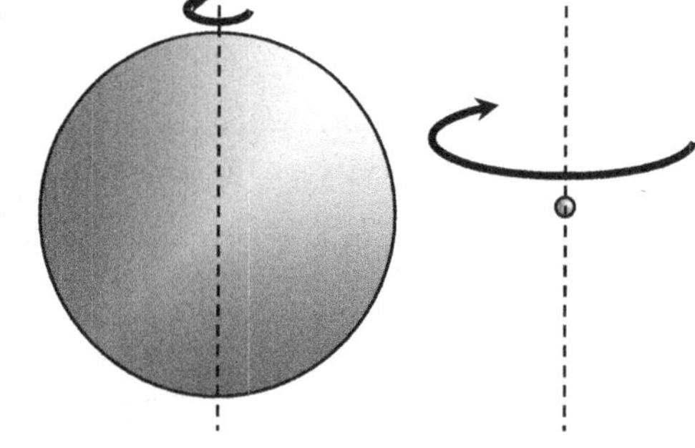

11.6 Formation of a neutron star: initially star rotates with period of 30 days. The star collapses from radius of 10^4 km to 3 km. Note: figure is not to scale; the final star is actually **<u>much</u> <u>smaller</u>** than indicated in the figure!

a) Write down an equation that relates period to angular frequency.
b) Write an expression for the conservation of angular momentum in terms of *period* instead of angular velocity.

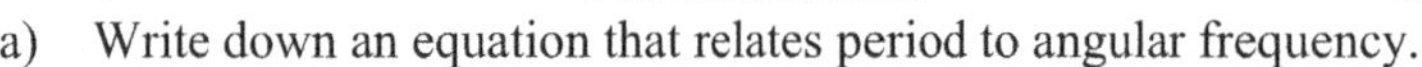

c) Determine new period.
d) Determine the *fractional* energy change.
   Fractional energy change is $\Delta E = \frac{E_f - E_i}{E_i}$.
e) What force did work to change the energy of the star?

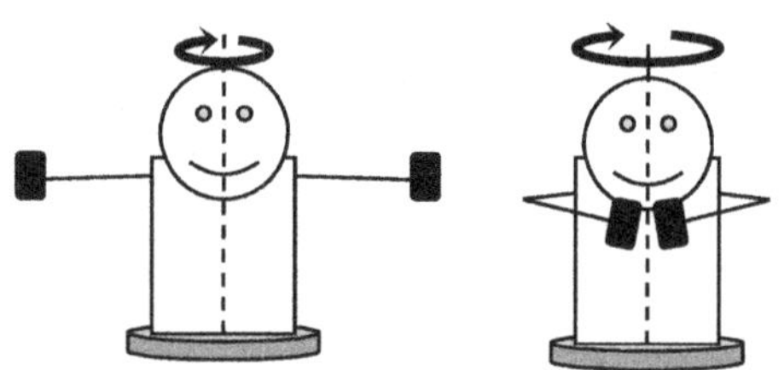

11.7 A class volunteer weighing about 154-lbs (70 kg) stands on the rotation platform. She or he holds in *each* hand a 10-lbs (4.5 kg) weight. To simplify the calculations, assume with arms out the moment of inertia is about 8 $\text{kg}\cdot\text{m}^2$. With her arms in assume the moment of inertia drops by about 25%.

a) Assume the initial period of rotation is 1 sec.
   Determine the period of rotation after the volunteer brings her or his arms in.
b) **Challenge:** do measurements of your own body and try to calculate the moment of inertia with arms out and arms in. I would assume the arms are uniform rods, the body/legs could be flat plate (or cylinder), and extra weights are point masses. Don't overthink it or try to get to many sig figs. It is just an approximation. I made my own attempt in the solutions.

HEY YOU! DO NOT USE CONSERVATION OF ENERGY ON A CONSERVATION OF ANGULAR MOMENTUM PROBLEM. HAVE YOU NOTICED THAT ENERGY IS NOT CONSERVED ON THE LAST FEW PROBLEMS?!?!? Don't say I didn't warn you when you get your test back.

11.8 This one has a ton of tricky details. Good problem but very challenging to avoid typos.
A lump of clay with mass m moves with initial speed v towards a rod as shown in the figure. The rod is hanging freely connected to a frictionless pivot. The clay collides and sticks to the end of a rod (mass $2m$ and length d). The collision time is very short. Ultimately the rod-clay system swings up to some maximum angle θ from the vertical.

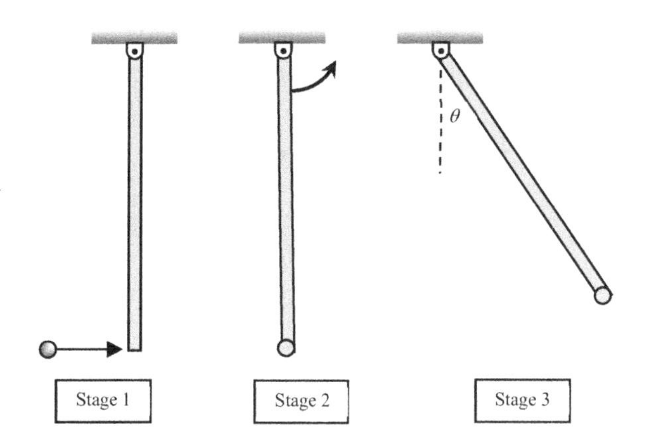

Goal: determine the final angle θ in terms of d, v, and g.

a) Between which stages is it appropriate to apply conservation of angular momentum? Circle all that apply.

Between 1 & 2	Between 1 & 3	Between 2 & 3	None of these	Impossible to determine

b) Explain the reasoning behind the previous answer.

c) Between which stages is it appropriate to apply conservation of energy? Circle all that apply.

Between 1 & 2	Between 1 & 3	Between 2 & 3	None of these	Impossible to determine

d) Explain the reasoning behind the previous answer.

e) Explain why the center of mass of the rod-clay relevant in this problem.

f) Show the max angle is given by

$$\theta_{max} = \cos^{-1}\left(1 - \frac{3v^2}{20gd}\right)$$

g) Show the system loses 40% of the initial energy during the collision.

11.9 In deep space a lump of clay with mass m moves with initial speed v towards a rod as shown. The rod has mass $2m$ and length d is initially at rest. The clay collides with and sticks to the end of the rod.

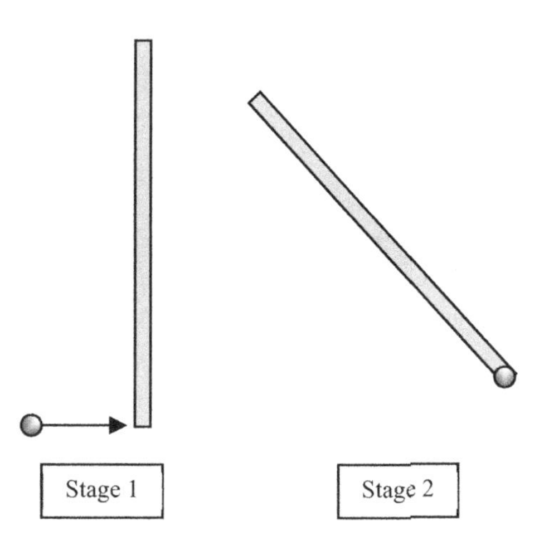

a) Which of the following conservations laws apply when comparing just before and just after the collision? Circle all that apply.

Energy Linear Momentum Angular Momentum

b) Explain why the center of mass is relevant in this problem.

c) Show that the moment of inertia of the rod-clay system about the center of mass of the system is given by $I = \frac{1}{3}md^2$.

d) Show the translational speed after the collision is $v_f = \frac{v}{3}$.

e) Show the final rotational speed about the center of mass of the rod-clay system is $\omega_f = \frac{v}{d}$.

f) Show the rod-clay system loses 33.3% of its initial energy in the course of the collision.

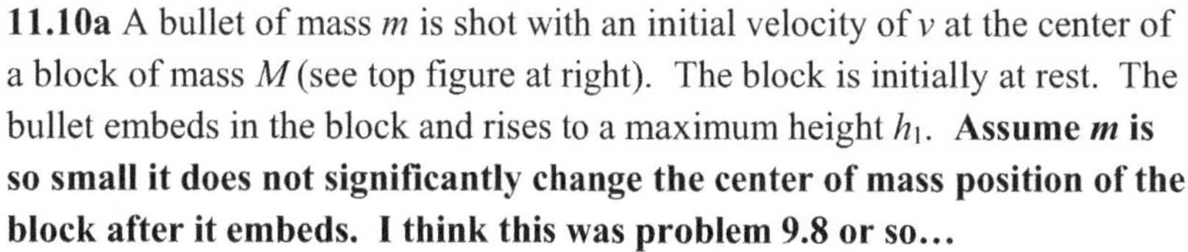

11.10a A bullet of mass m is shot with an initial velocity of v at the center of a block of mass M (see top figure at right). The block is initially at rest. The bullet embeds in the block and rises to a maximum height h_1. **Assume m is so small it does not significantly change the center of mass position of the block after it embeds. I think this was problem 9.8 or so…**

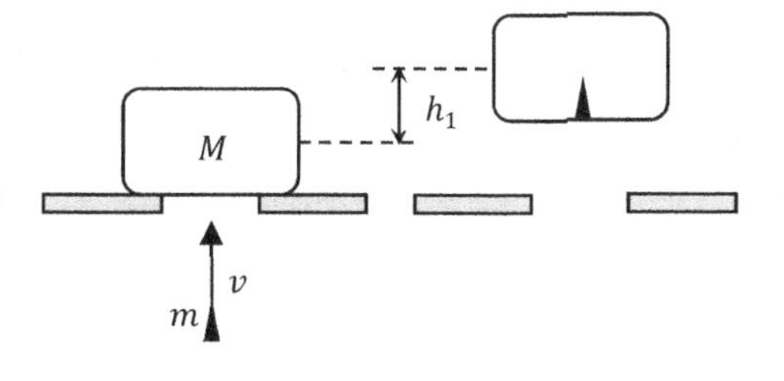

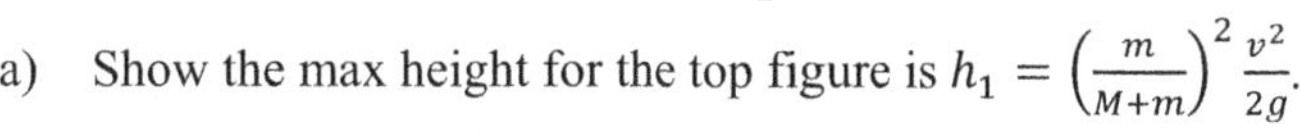

a) Show the max height for the top figure is $h_1 = \left(\frac{m}{M+m}\right)^2 \frac{v^2}{2g}$.

b) Show the energy change during the collision is

$$\%\Delta E = \left(\frac{m}{m+M} - 1\right) \times 100\% = \frac{-M}{m+M} \times 100\%$$

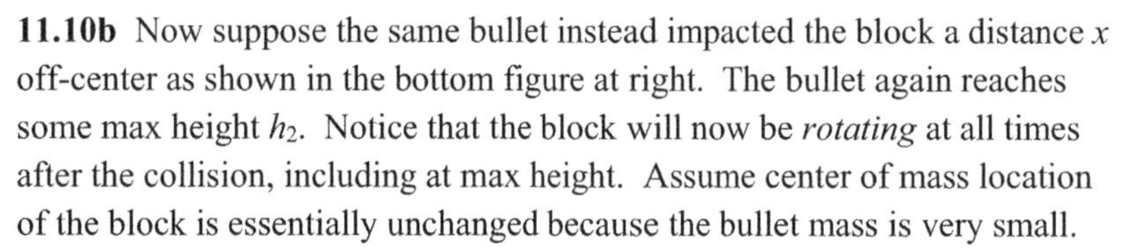

11.10b Now suppose the same bullet instead impacted the block a distance x off-center as shown in the bottom figure at right. The bullet again reaches some max height h_2. Notice that the block will now be *rotating* at all times after the collision, including at max height. Assume center of mass location of the block is essentially unchanged because the bullet mass is very small.

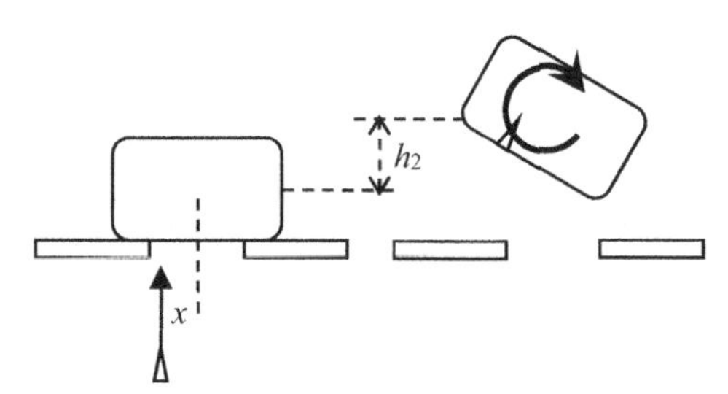

a) How does h_2 compare to h_1? Circle the best answer.

$h_2 > h_1$ $h_2 = h_1$ $h_2 < h_1$

b) Which collision should lose *less* energy? Why?

c) Show the energy change after the collision is

$$\%\Delta E = \left(\frac{mx^2}{I} + \frac{m}{m+M} - 1\right) \times 100\%$$

where I is the moment of inertia of block and bullet after the collision.

Deep Thoughts

In free space an object tends to rotate about its system center of mass.
In free space one uses both linear and angular momentum during the collision.
When a pivot is located on the object, it will rotate about that pivot.
For an object swinging in a circular arc, gravitational potential energy relates to the center of mass radius.
Collisions involving rotation and translation lose less energy when compared to similar translation-only collisions.

11.11 A space probe is essentially a solid disk of radius R with a retractable antenna shaped like a thin rod of length $2R$. Initially the antenna is retracted, the ends of the antenna line up with the disk as shown in the figure. The probe is slowly rotating at 2.80 RPM about the dotted line axis shown in the figure. The probe then extends its antenna to signal earth. When the antenna is fully extended, the edge of the antenna is essentially at the edge of the disk. The mass m of the antenna is equal to the mass of the rest of probe (the disk).

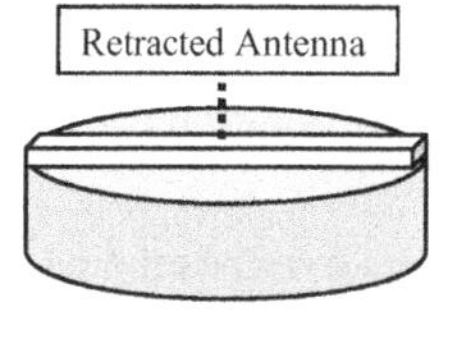

Extended Antenna

a) Should you do an energy problem or angular momentum problem to analyze this system before and after the antenna is extended? Explain why?
b) What is the initial moment of inertia (before the antenna has extended)?
c) Where is the center of mass after the antenna is extended?
d) What is the final moment of inertia (after the antenna has extended)?
e) What is the new rotation rate (in RPM) of the probe after the antenna is extended?
f) What is the % change in rotational kinetic energy? What does work to change the energy?
g) How does the period of the probe's rotation change? Circle the best answer.

Increases | Decreases | Stays the same | Impossible to determine

h) How does the probe's angular momentum change? Circle the best answer.

Increases | Decreases | Stays the same | Impossible to determine

i) How does the probe's rotational kinetic energy change? Circle the best answer.

Increases | Decreases | Stays the same | Impossible to determine

j) How does the probe's translational kinetic energy change? Circle the best answer.

Increases | Decreases | Stays the same | Impossible to determine

k) After the probe is extended where is the axis of rotation?

At the center of the disc | At the center of mass | Both previous answers are correct | None of the previous answers are correct

11.11$\frac{1}{11}$ (11.Nasty) A bullet of unknown mass travels with speed v just before impact with a wooden plate (mass m and side length s). After impact, the bullet embeds in the top left corner of the plate. There is a small wedge affixed to the ground beside the bottom right corner of the square plate. Determine the minimum bullet mass required (with the given initial speed v) to cause the plate to tip over and land on its side. You may assume the bottom right corner of the plate experiences negligible slipping with the ground as the plate tips over.

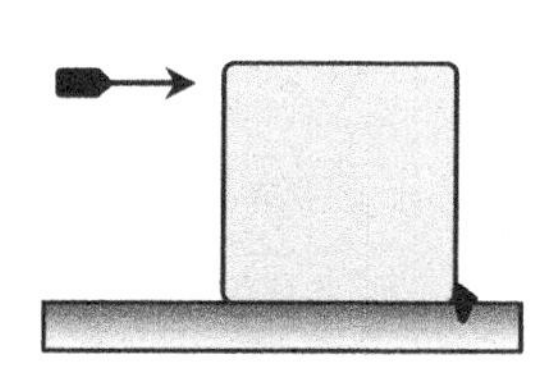

Note: I labeled the unknown bullet mass m_1 in the solutions. If you get stuck, the first page of the solution has a hint. Figuring out what equations to use and getting this problem started is *outstanding* practice. If you can get an equation with all givens except m_1 you are doing great work!

11.12 Spinning wheel on turntable

Consider a person standing on a turntable. The turntable has excellent bearings (negligible axle friction). The person and the turntable have no net external force (gravitational force exerted by earth is balanced by the normal force exerted by the earth). Remember, n and mg are NEVER an action-reaction pair.

The person is holding a wheel. A second person, not shown in the figure, grabs the first person and holds them stationary (prevents them from twisting around). While holding the 1st person, the second person spins the wheel as quickly as possible. The 2nd person then releases the first person. At this point the first person is completely stationary holding the spinning wheel as shown in the "*B4* Flipping Wheel" figures below. Note: in the top view assume axis of rotation is out of the page as indicated by the figure.

Then the person twists the wheel by 90°. The direction of *wheel* rotation is shown in the two figures labeled "*After* Flipping Wheel".

For this one, please attempt all parts before looking at the answers…

Predict which direction the person will rotate and sketch that in the figures below.

Which of the following is/are conserved in this demonstration: energy, angular momentum, translational momentum, force, Chewbacca-ness? More than one answer may be possible.

Think: the wheel's rotation *rate* does not change significantly when you twist it (obviously the rotation direction does change). The human's rotation rate increases. Is energy conserved? Feel free to reconsider your answers to the previous question. Also think, where does this extra energy come from?

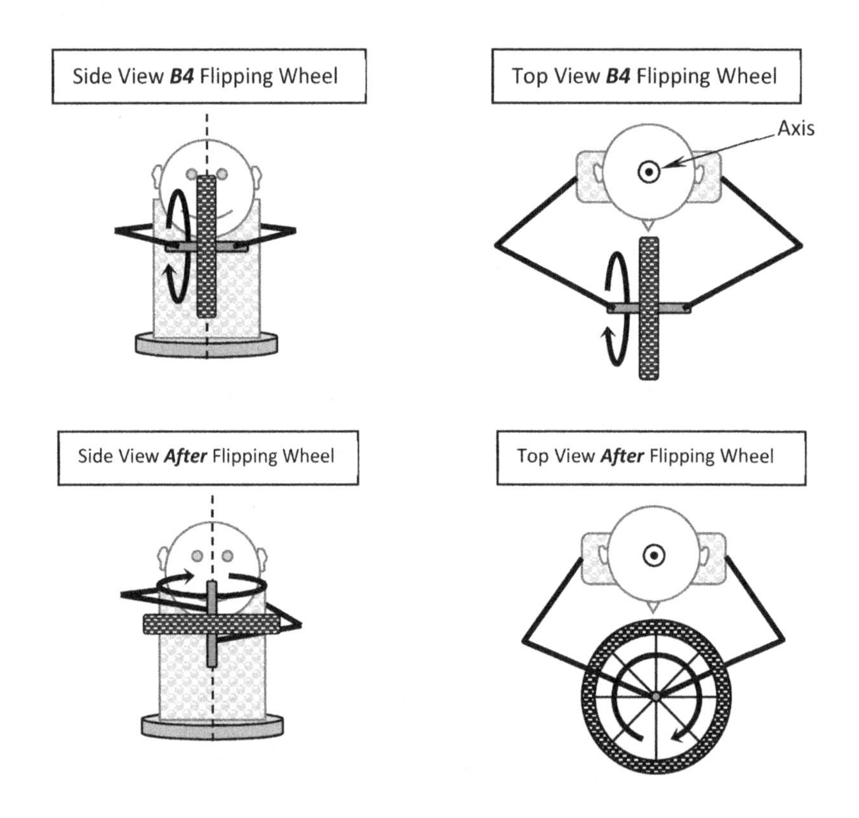

11.13 Magic Wheel Isn't Magic

A person tries to hold a wheel when it is NOT spinning. They grab the very end of the axle on only one side. It is extremely difficult if not impossible for most people to hold the wheel this way.

If the same person tries to hold the spinning wheel while it IS spinning it feels much easier. The weight of the wheel has not changed. Spinning the wheel does nothing to reduce the gravitational attraction the earth exerts on the wheel. Why is it easier to support the wheel?

Side View ***After*** Spinning Wheel

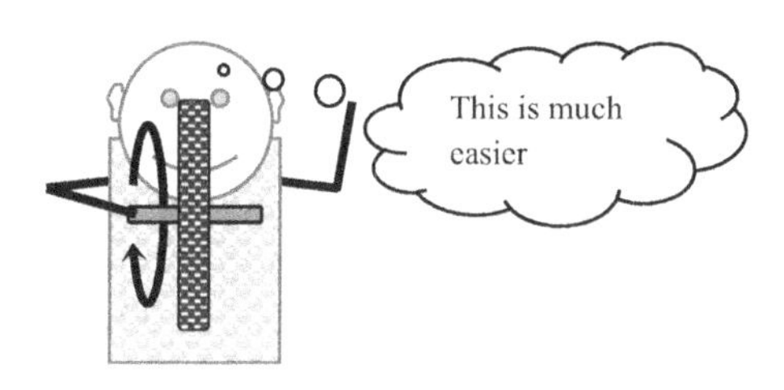

When your hand grabs the wheel while it is not spinning, your hand is actually exerting multiple forces to balance the torques. The part of your hand closest to the center of the wheel becomes the pivot point and exerts an upwards force. The part of your hand at the very end of the axle must exert a large downwards force to balance the torque of the wheel's weight. As a result, the part of your hand closest to the wheel must exert a force larger than the wheel's weight to balance the system. Consider a web search for "Anti-gravity wheel" to see a very well made video...

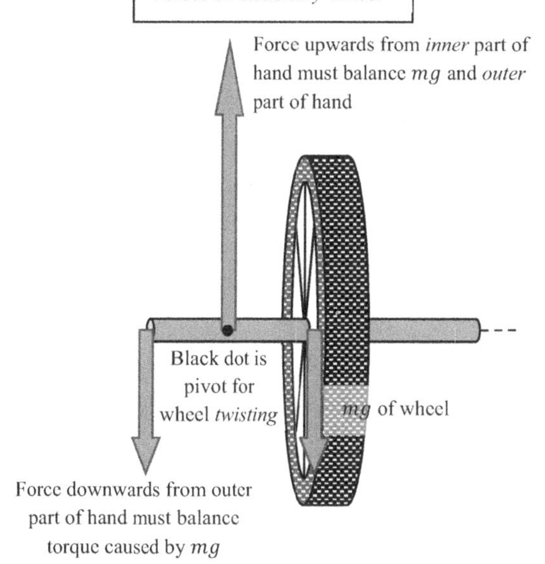

Once the wheel is spinning try supporting it at the very end with a rope or, if you are fairly strong, with your little finger. You will immediately notice you can support the weight but the wheel begins *precessing* while it spins (but does not twist). The *precessional* motion of the *spinning* wheel can be explained by the following:

$$\vec{\tau}_{ext} = \frac{d\vec{L}}{dt}$$

$$(y\hat{\jmath}) \times (-mg\hat{k}) = \frac{d\vec{L}}{dt}$$

$$mgy(\hat{\jmath}) \times (-\hat{k}) = \frac{d\vec{L}}{dt}$$

$$mgy(-\hat{\imath}) = \frac{d\vec{L}}{dt}$$

Notice the spinning wheel has angular momentum $\vec{L}$. I assumed the length of the axle rod held by the hand is y. The torque caused by the weight of the wheel changes the angular momentum of the wheel into the page.

Challenge: Try doing the experiment twice; once spinning clockwise and once spinning counterclockwise. Is the direction of precessional motion affected by the direction of wheel rotation?

Forces on *Spinning* Wheel

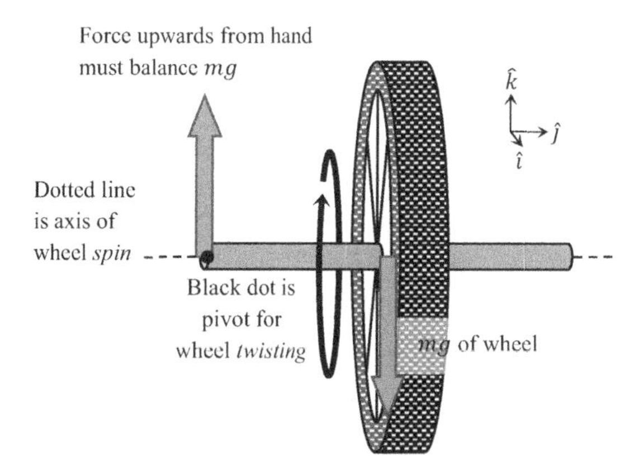

11.14 Secret Suitcase

A spy leaves suitcase unattended in the airport. You detect the unattended suitcase and you pick it up, run forwards a few steps, then turn 90° while holding the suitcase so you can throw it down an empty hallway (just in case it is full of explosives…or stinky anchovy paste to be used in a synthetic cheese dip product).

What the heck is going on with this crazy suitcase?

11.15 Riding a bike

It is easier to maintain balance on a bike while you are riding as opposed to when you are stationary. Explain why.

11.16 A person of mass M stands at the center of a rotating turntable. The turntable is a disk with the same mass M and radius R and rotates with an initial rate of ω_0. The person walks directly towards the edge of the turntable (second figure). At the instant shown in the second figure, the person is distance r from the center. To simplify the problem, assume the person is as a point mass. A point mass model for the person model works well whenever $d \ll R$…

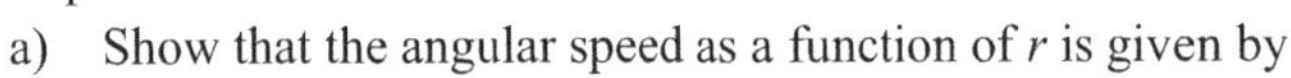

a) Show that the angular speed as a function of r is given by

$$\omega = \omega_0 \frac{R^2}{2r^2 + R^2}$$

b) Sketch a plot of ω vs r. Ignore units. Assume $\omega_0 = R = 1$.

c) **Requires Calculus** Determine maximum rotational energy *of the person* moving from $r = 0$ to $r = R$.

d) Sketch a plot of the person's rotation energy versus r. Ignore units. Assume $M = \omega_0 = R = 1$.

e) **Requires Calculus** Assume the person starts at time $t = 0$ and moves at a constant rate v towards the edge of the disk. Show the *magnitude* of angular acceleration of the disk is given by

$$\alpha = \frac{4\omega_0 R^2 v^2 t}{(2v^2t^2 + R^2)^2}$$

and the angular displacement of the disk is given by

$$\Delta\theta = \omega_0 t - \frac{\omega_0 R}{\sqrt{2}v}\tan^{-1}\left(\frac{\sqrt{2}vt}{R}\right)$$

f) **Going further:** You could assume the person is actually a cylinder of diameter d. At the center the moment of inertia is thus $\frac{1}{8}Md^2$. As the person moves out you could use the parallel axis theorem to find the person's moment of inertia as $\frac{1}{8}Md^2 + Mr^2$. I haven't done it, but it might be interesting to see if this changes things significantly for small radii. I would make reasonable assumptions for the masses and radii and create plots of ω vs r for each scenario to get a feeling for the differences in the two models.

11.17 Relative velocity challenge A person of mass M stands on the edge a stationary turntable of mass $m = \frac{M}{3}$. The turntable is essentially a disk with radius R. The person begins to walk with a constant speed of v *relative to the disk* (not relative to the earth!) as shown in the 2nd figure.

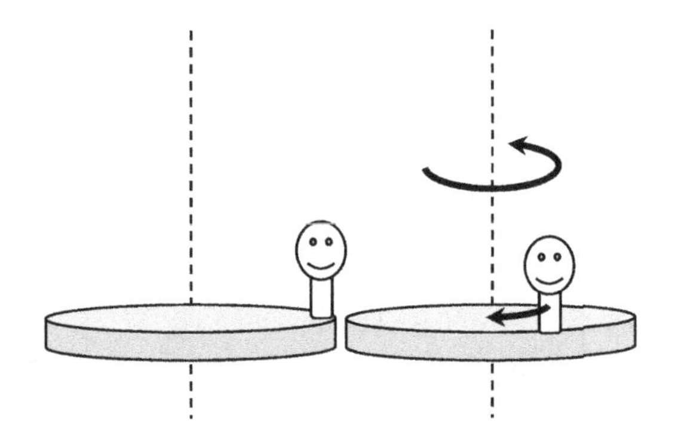

a) Show the rotation rate of the turntable *relative to the earth* is given by $\frac{6v}{7R}$. Hint: consider the zombie walking on the board problems late in chapter 9!

b) Show that the rotation rate of the person *relative to the earth* is given by $\frac{v}{7R}$. Watch your +/- signs!

The Big Three Conservation Laws of Mechanics

Physicists love to use conservations laws. From a practical standpoint, conservation laws allow physicists to write down an equation that relates a before picture to an after picture. Properly identifying which conservation law or laws apply to a physical situation is crucial. Many physicists would argue the biggest takeaway from this course should be a thorough understanding of these laws *and* how to apply them. To that end, the point of these questions is to see if you know *when to use the laws* more than the technical details of executing the calculations.
Random note: the 2nd two laws relate to the quantum numbers of the periodic table.

Special Condition	Equation	Conservation Law
if $\Sigma\vec{F}_{ext} = 0$	$\vec{p}_i = \vec{p}_f$	Conservation of Momentum
if $\Sigma\vec{\tau}_{ext} = 0$	$\vec{L}_i = \vec{L}_f$	Conservation of Angular Momentum
$W_{ext/non-conervative} = 0$	$E_i = E_f$	Conservation of Energy

11.18 Suppose a lump of clay of mass m is thrown at the center of a stationary rod of mass m and length d deep in outer space far from any other masses. The clay sticks to the rod after the collision.

a) During the collision, which conservation law or laws, if any, apply? Assume the system includes the rod and the clay.
b) Draw a picture of the rod (with clay stuck to it) after the collision. In particular, is the rod translating, rotating, or both?
c) If the rod is rotating after the collision, about what point is the rod rotating: the end of the rod, the center of the rod, or somewhere else?

Before

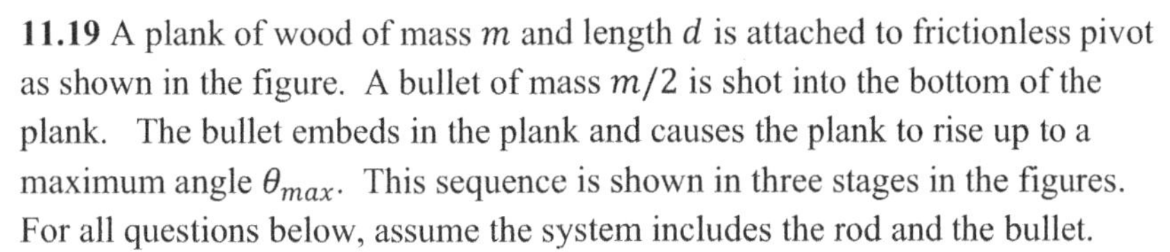

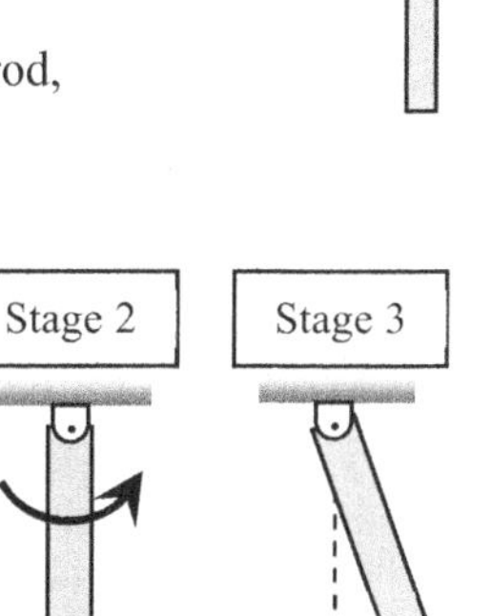

11.19 A plank of wood of mass m and length d is attached to frictionless pivot as shown in the figure. A bullet of mass $m/2$ is shot into the bottom of the plank. The bullet embeds in the plank and causes the plank to rise up to a maximum angle θ_{max}. This sequence is shown in three stages in the figures. For all questions below, assume the system includes the rod and the bullet.

a) Between Stage 1 and Stage 2, which conservation laws, if any, apply?
b) Between Stage 2 and Stage 3, which conservation laws, if any, apply?
c) Between Stage 1 and Stage 3, which conservation laws, if any, apply?
d) What is the moment of inertia of the plank with bullet in it?
e) What is the distance from the pivot to the center of mass after the collision?
f) What is the initial speed of the bullet? Answer in terms of g, L, and θ_{max}.

11.20 A steel sphere of mass m is thrown at the end of a stationary steel rod deep in outer space far from any other masses. The rod has mass $2m$ and length d. The collision is essentially elastic.

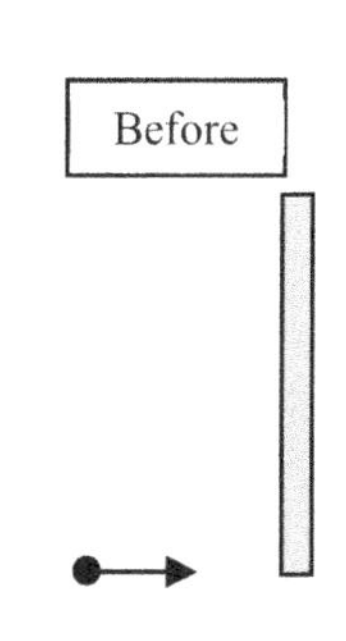

a) During the collision, which conservation law or laws, if any, apply? Assume the system includes the rod and the clay.
b) Draw a picture after the collision. In particular, is the rod translating, rotating, or both?
c) If the rod is rotating after the collision, about what point is the rod rotating: the end of the rod, the center of the rod, or somewhere else?
d) What is the moment of inertia of the rod while it is rotating?

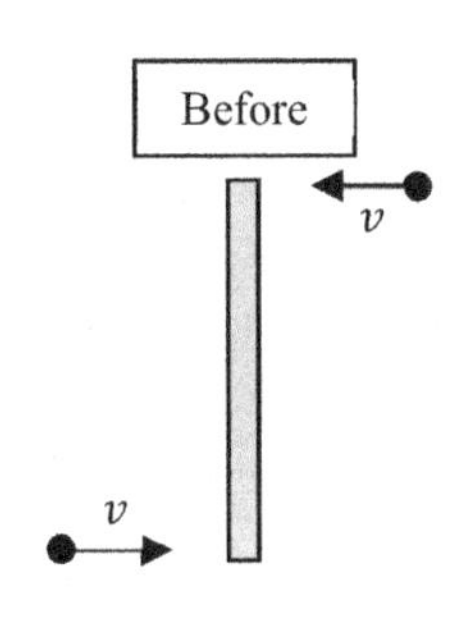

11.21 Two lumps of clay each have mass $m/2$ and are moving in opposite directions with the same speed as shown. The lumps impact a stationary rod of mass m and length d deep in outer space far from any other masses. The lumps of clay simultaneously impact, and stick to, the ends of the rod.

a) During the collision, which conservation law or laws, if any, apply? Assume the system includes the rod and the clay.
b) Draw a picture of the rod (with clay stuck to it) after the collision. In particular, is the rod translating, rotating, or both?
c) If the rod is rotating after the collision, about what point is the rod rotating: the end of the rod, the center of the rod, or somewhere else?
d) What is the moment of inertia of the rod once the clay is attached? You may assume each lump of clay is a point mass.

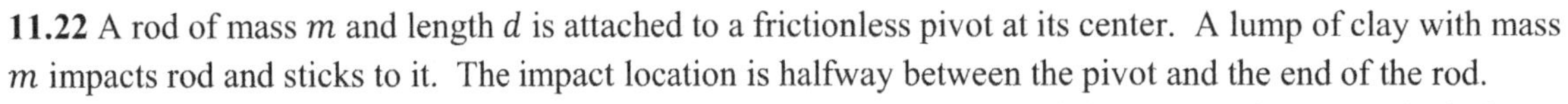

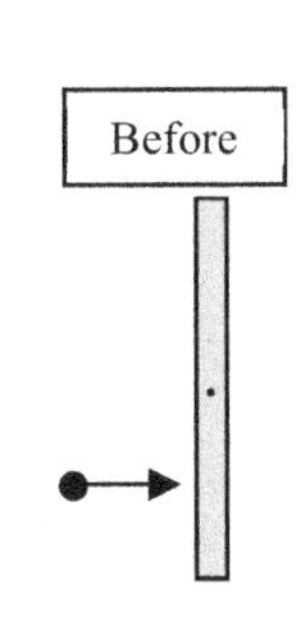

11.22 A rod of mass m and length d is attached to a frictionless pivot at its center. A lump of clay with mass m impacts rod and sticks to it. The impact location is halfway between the pivot and the end of the rod.

a) During the collision, which conservation law or laws, if any, apply? Assume the system includes the rod and the clay.
b) Draw a picture of the rod (with clay stuck to it) after the collision. In particular, is the rod translating, rotating, or both?
c) If the rod is rotating after the collision, about what point is the rod rotating: the end of the rod, the center of the rod, or somewhere else?
d) What is the moment of inertia of the rod once the clay is attached? You may assume the lump of clay is a point mass.

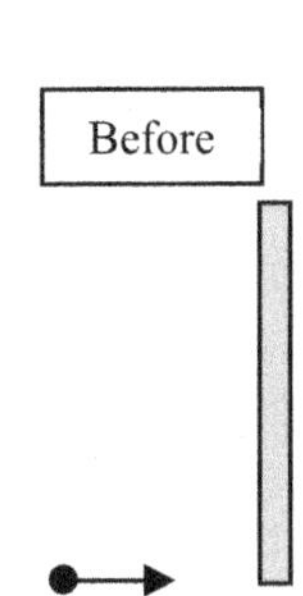

11.23 Suppose a lump of clay with mass m is thrown at the end of a stationary rod deep in outer space far from any other masses. The rod has mass $3m$ and length d. The clay sticks to the rod after the collision.

a) During the collision, which conservation law or laws, if any, apply? Assume the system includes the rod and the clay.
b) Draw a picture of the rod (with clay stuck to it) after the collision. In particular, is the rod translating, rotating, or both?
c) If the rod is rotating after the collision, about what point is the rod rotating: the end of the rod, the center of the rod, or somewhere else?
d) What is the moment of inertia of the rod once the clay is attached? You may assume the lump of clay is a point mass.

Before

11.24 Suppose a lump of clay with mass m is thrown at the end of a moving rod deep in outer space far from any other masses. The rod has mass m and length d. The rod and the clay move with equal speeds but in opposite directions. The clay sticks to the rod after the collision.

a) During the collision, which conservation law or laws, if any, apply? Assume the system includes the rod and the clay.
b) Draw a picture of the rod (with clay stuck to it) after the collision. In particular, is the rod translating, rotating, or both?
c) If the rod is rotating after the collision, about what point is the rod rotating: the end of the rod, the center of the rod, or somewhere else?
d) What is the moment of inertia of the rod once the clay is attached? You may assume the lump of clay is a point mass.

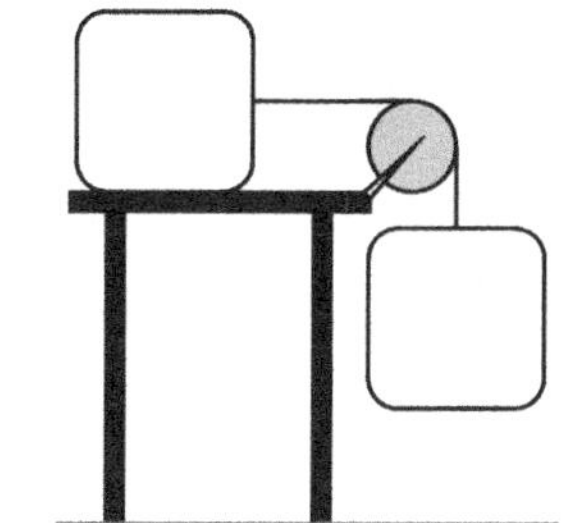

11.25 Consider the Fletcher's trolley shown at right. Assume both blocks have mass m and the pulley is a disk with radius R and mass m. Assume the interface between the block and the table has negligible friction. The system is released from rest. Just before reaching the ground the pulley is rotating with rate ω. Assume your system includes only the blocks, the pulley and the string connecting them. To be clear, the earth is not included in the system.

a) Which conservation laws are valid while the hanging mass falls towards the floor?
b) If the conservation laws are violated, there must be some external work, force, or torque. For each violated conservation law, explain the external work, force, or torque which invalidates the use of the conservation law.
c) If the earth is included in the system, what happens to your answer to part a?
d) Why does it appear that the conservation laws are violated even when the earth is included in the system? For instance, initially the system is at rest and thus has no angular momentum. Once falling, the blocks and pulley have angular momentum. Explain why conservation of angular momentum is not actually violated even though $\vec{L}_i \neq \vec{L}_f$ for the blocks and pulley.

11.26 A figure skater is spinning on ice. He brings his arms in. His rotation rate increases after he brings his arms to his chest. Typically this problem is analyzed using conservation of angular momentum…not conservation of energy.

a) Would you expect moment of inertia to increase, decrease, or remain constant?
b) Would you expect rotation rate to increase, decrease, or remain constant?
c) Would you expect rotational kinetic energy to increase, decrease, or remain constant?
d) Explain why conservation of energy is not valid for the situation described.

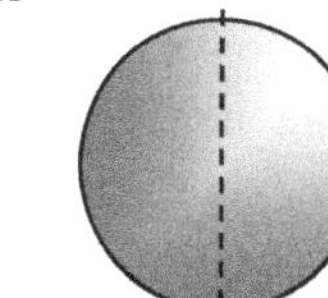

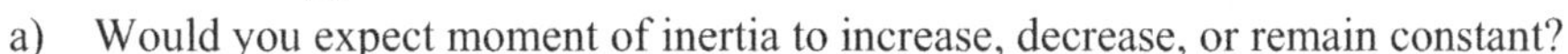

11.27 A star is initially rotating with rate ω_0. It collapses to form a neutron star of much smaller radius. Typically this problem analyzed using conservation of angular momentum…not conservation of energy.

a) Would you expect moment of inertia to increase, decrease, or remain constant?
b) Would you expect rotation rate to increase, decrease, or remain constant?
c) Would you expect rotational kinetic energy to increase, decrease, or remain constant?
d) Gravity, the force causing the star to collapse, is a conservative force. This implies one can use conservation of energy do determine the change in rotation rate. Why is this not a straightforward calculation if one uses conservation of energy?

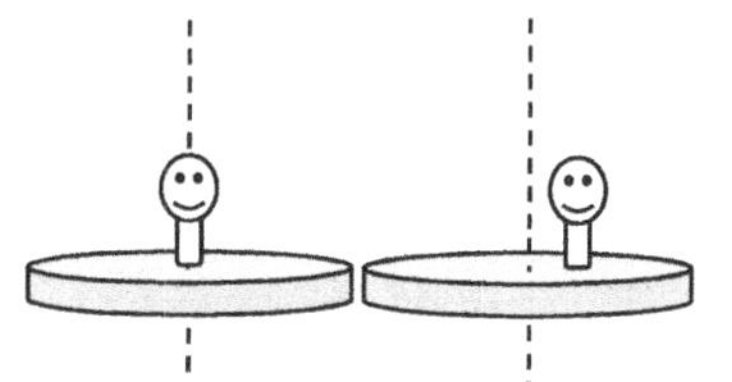

11.28 A zombie stands at the center of a large rotating disk. The zombie walks radially outwards towards the edge of the disk.

a) Would you expect moment of inertia to increase, decrease, or remain constant?
b) Would you expect rotation rate to increase, decrease, or remain constant?
c) Would you expect rotational kinetic energy to increase, decrease, or remain constant?
d) Explain why conservation of energy is not valid for the situation described.

Notice in systems that deform (change shape) conservation of energy is not the recommended. Some form of work is being done.

- In problems **11.26** & **11.28**, work is done by non-mechanical means (conversion of chemical potential energy). I have no idea how to compute this amount of work.
- In problem **11.27**, work is done by gravity but we are no longer dealing with a point mass near the earth's surface. It is a non-trivial task to determine the amount of gravitational potential energy converted to kinetic energy.

For objects that rotate and change shape, use conservation of angular momentum.

For collisions between a rod and a point mass sometimes momentum is not conserved. Unless the collision is elastic, one expects energy is <u>not</u> conserved.

If the rod is attached to a frictionless pivot:

- The pivot will have a reaction force which exerts an external force. Conservation of momentum <u>does not</u> apply.
- Since the reaction force acts at the pivot, the reaction force will not cause an external torque. Conservation of angular momentum <u>does</u> apply.

If the rod is freely floating in space:

- No external forces or torques are present during the collision. Both momentum and angular momentum are conserved.
- After the collision, objects will rotate about the center of mass. If a piece of clay sticks to the rod, find the center of mass of the rod-clay system and use that as the pivot point for the problem!

11.29 Two blocks of mass m and $2m$ are initially at rest on a frictionless horizontal surface. A person, not shown in the figure, is initially compressing a light spring of constant k between the blocks. Upon releasing the blocks, the spring falls to the ground between the blocks and is no longer touching either block. After release, mass m travels to the right with speed v.

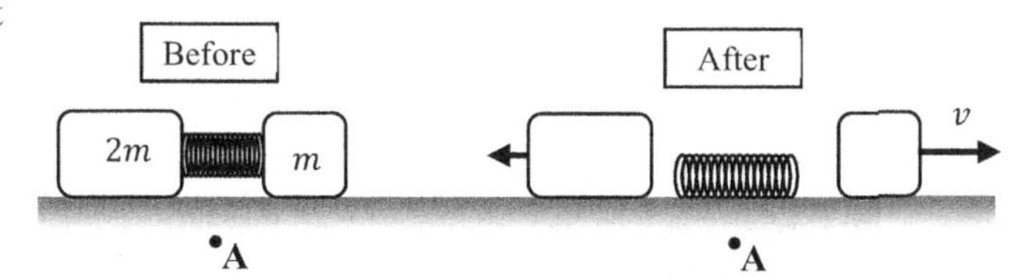

a) During the situation described, which conservation law or laws, if any, apply? Assume the system includes the blocks and the spring.

b) Consider v_{2f}, the final speed of mass $2m$, and the initial compression of the spring (x).
Since there are only two unknowns, only two conservation laws need be applied.
Traditionally this problem is solved using conservation of momentum and conservation of energy.
Could it be solved using conservation of angular momentum and conservation of energy?
Explain why or why not.
Hint: consider the angular momentum about point **A** (or some other point not in the line of motion).

11.30 Consider a moon in a circular orbit of radius *r* about a planet. The moon has mass *m* and speed *v*. The moon is the small grey circle on the right side of the orbit. No mass is present at point **A**; it is merely a point labeled for ease of communication. The planet is centered on the origin labeled **B**. As the moon orbits the planet, you may assume the mass of the moon is very small compared to the planet. The center of mass of the moon-planet system is essentially at point **B**. The only force acting on the moon is a force of gravitational attraction towards the planet.

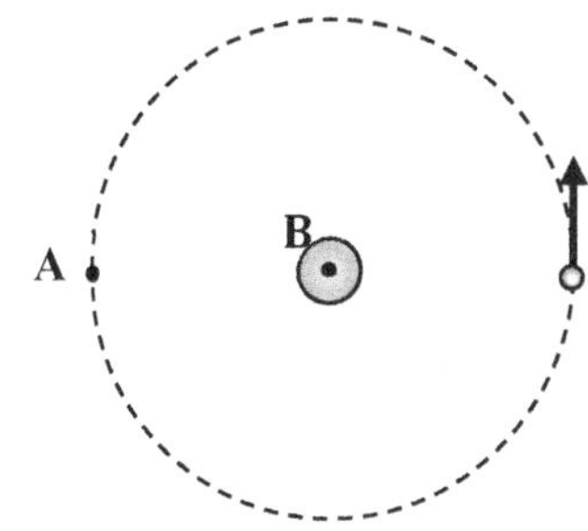

a) Is the angular momentum of the moon about point **B** conserved? Explain why or why not. Hint: consider the next external torque acting on the moon if point **B** is the pivot.
b) Is the angular momentum of the moon about point **A** conserved? Explain why or why not. Hint: consider the next external torque acting on the moon if point **A** is the pivot.
c) Determine the angular momentum of the moon at the center of the orbit (about point **B**).

11.31 A simple pendulum is constructed from a lump of clay of mass m and light string of length d as shown in the figure. A bullet of mass $m/2$ is shot into the clay. The bullet embeds in the clay and causes it to rise up to a maximum angle θ_{max}. This sequence is shown in three stages in the figures. For all questions below, assume the system includes the clay and the bullet.

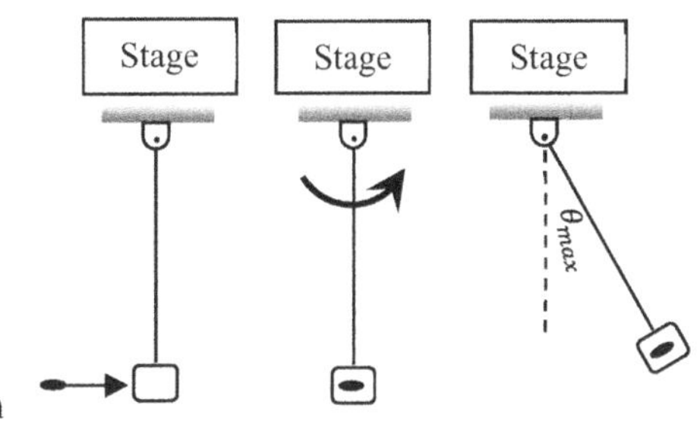

Also notice, this is a variation of question 2 where we have replace the rod (and extended object) with a point mass. Note: in question 2 we said between stage 1 and stage 2 $\vec{L}_i = \vec{L}_f$ but $\vec{p}_i \neq \vec{p}_f$.

a) Determine the max angle using $\vec{L}_i = \vec{L}_f$.
b) Determine the max angle using $\vec{p}_i = \vec{p}_f$.

Explain why $\vec{p}_i = \vec{p}_f$ is valid in this case but not problem 2!

11.32 This is a messier version of problem 4. Two lumps of clay each have mass $m/2$ and are moving in opposite directions (same speed v) as shown. The lumps impact a stationary rod of mass m and length d deep in outer space far from any other masses. The lumps of clay simultaneously impact, and stick to, the end and center of the rod.

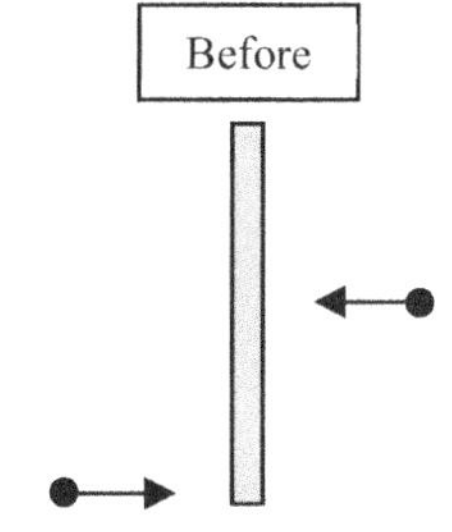

a) Determine the translational velocity of the rod after the collision.
b) Determine the rotational velocity of the rod after the collision.

If you are feeling bold, try to code a simulation of the above experiment.

- Allow the user to select all masses and the impact locations.
- Have your code use conservation of *linear* momentum to determine the *translational* velocity after the collision.
- Have your code compute the center of mass position.
- Have your code use conservation of *angular* momentum to determine the *rotational* velocity after the collision.
- Update the position using $\Delta\vec{x}_{CM} = \vec{v}_{after\ collision}\Delta t$
- Update the *angular* position using $\Delta\vec{\theta} = \vec{\omega}_{after\ collision}\Delta t$
- **Strictly speaking, we may only treat $\Delta\vec{\theta}$ as a vector for rotation in a single plane.**

STATIC EQUILIBRIUM

Whenever an object is 1) not rotating and 2) not translating we say it is in static equilibrium.

- The lack of *rotation* implies the sum of *torques* is zero ($\Sigma\vec{\tau} = 0$) about any point.
- The lack of *translation* implies the sum of *forces* is zero ($\Sigma\vec{F} = 0$).

1) Draw a picture and get a feel for the problem.
   a) Try to draw all the forces on the problem.
   b) It is often useful to draw a *separate* sketch showing only lengths & angles without any forces.
   c) If there is an actual pivot in the problem don't forget that there are *reaction* forces in the problem! The reaction forces are usually written R_x & R_y and act at the physical pivot.
2) Do the sum of torques first.
   a) Choose a pivot point. You are free to choose your pivot point or axis in static equilibrium problems since if the object in the problem is not rotating, it is not rotating about ANY pivot point.
   b) Figure out which forces cause no torques (because line of action runs through pivot point).
   c) Sketch the problem drawing only those forces that will cause torques. Clearly show the pivot point. Choose a direction to call positive for the entire problem (usually either CW or CCW).
   d) List out remaining torques one at a time and sketch the *r*, *F*, and angle for each torque.
      i) Double check the angles for each force.
      ii) Verify you correctly drew $\vec{r}$ from the pivot to the point where force is applied.
      iii) Check the sign according to your choice of coordinates.
      iv) If possible, write r in terms of L (say $L/2$ or $L/6$)
      v) Use whatever method works best for you ...$\vec{\tau} = \vec{r} \times \vec{F} = rF\sin\theta = r_\perp F = rF_\perp$
   e) Finally write down the sum of torques eqt'n keeping track of all the +/- signs. Check if L drops out...
3) Now do a sum of force problem from scratch
   a) Start a new picture of the object for your FBD...I know, it's your third or fourth one at this point.
   b) On this FBD it is usually best to use the standard xy coordinate system.
   c) Now you need to split up the forces into x and y components. This might look similar to (or totally different from) your torque diagram. **Do not assume the forces split up the same way as in your diagram for the torques!**
   d) Do the standard force equation crap.
4) Now you should probably have three equations. One for the torques and probably a couple from the forces.
   a) List them neatly in one place.
   b) Re-read the problem to determine WTF. Do you have enough equations and to handle the number of unknowns in the problem? If not try doing the sum of torques using a different pivot point.
5) Use your equations to solve for WTF. Remember, if "just about to slip" you can use the fact that $f = \mu_s n$. If the problem has friction but is not "just about to slip" assume *f* is some variable and DO NOT use $\mu_s n$!
6) Check your answers. If for some reason you get a negative number for a tension or a mass you did something seriously wrong...start over. If you get a negative number for R_x or R_y that is actually ok.
   a) Suppose you drew R_x pointing to the right. After doing all the math you found $R_x = -3.0$ kN. This implies R_x has *magnitude* of 3.0 kN but actually points LEFT because $-3.0 \text{ kN } \hat{\imath} = +3.0 \text{ kN } (-\hat{\imath})$.
   b) If R_x or R_y is negative, your *magnitude* is correct but the *direction* is opposite the direction drawn.

In summary: For cases of static equilibrium both the sum of forces is zero and sum of torques is zero. The object is not rotating it is not rotating ABOUT ANY POINT. That means any and every point you choose can be a pivot point for the problem. Again, this only works if nothing is rotating. Also, as long as no acceleration we can use the fact that the tension in strings will be mg. Also, if an object rests on a horizontal plank it will exert a force of mg on the plank. AGAIN, THIS IS TRUE ONLY WHEN ACCELERATION IS ZERO.

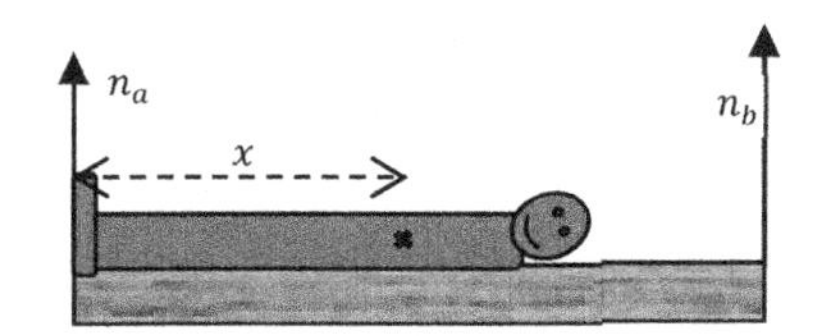

12.1 A plank of known length L is supported by a scale at each end. After the plank is placed on the scales the scales are zeroed. This is done to eliminate the need to consider the mass of the plank for the rest of the problem. A person lays on the plank as shown in the figure. The center of mass of the person is an unknown distance x from the left end of the plank.

a) Since the plank is not rotating, we know the sum of torques equals zero about point **A** (left end of plank). Write the torque equation for point **A** (left end of rod).
b) Write down the vertical force equation.
c) Determine the position x of the zombie's center of mass. In terms of the scale readings.
d) Check your results using the sum of torques about point **B** (right end of plank). Note: if the plank isn't rotating at all, we know it isn't rotating about ANY pivot point.

12.1½ Consider a rod mounted to a wall using a pivot with negligible friction. The rod has length L and mass m. A support cable with equal length (but negligible mass is used to support the rod). The support cable can be connected to the rod at any distance x from the pivot as shown. Mass $M \neq m$ hangs on a string from the end of the rod. Assume x, L, m & M are givens (as is the magnitude of freefall acceleration g).

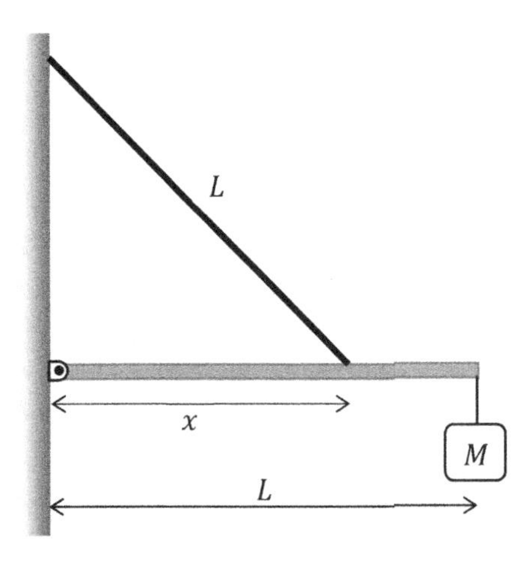

a) In general, tension in a string is *not always* equal to the weight of the mass hanging on it. Why is it ok, in *this* special case, to say the tension in the *string* (not the support cable) is Mg?
b) Write down sum of torques about the pivot.
c) Write down sum of forces in the x-direction using a standard coordinate system (positive x is to the right).
d) Write down sum of forces in the y-direction using a standard coordinate system (positive y is upwards).
e) Determine the tension *in the support cable* in terms of the givens. Express your answer as both a vector in Cartesian form ($\vec{T} = T_x\hat{\imath} + T_y\hat{\jmath}$) and in polar form (magnitude T & direction θ_T with sketch).
f) THINK: does your previous answer make sense when $x = 0$ or $x = L$? Consider the physical scenario for those two cases and try to imagine if tension should be large or small?
g) Determine the reaction force at the pivot in terms of the givens. Express your answer as both a vector in Cartesian form ($\vec{R} = R_x\hat{\imath} + R_y\hat{\jmath}$) and in polar form (magnitude R & direction θ_R with sketch).
h) THINK: does your previous answer make sense when $x = 0$ or $x = L$? Consider the physical scenario for those two cases and try to imagine if tension should be large or small?
i) Let $M = 4.00$ kg, $m = 1.00$ kg, $L = 1.00$ m. Create a spreadsheet tabulating T, θ_T, R & θ_R as $x = 0.05 \rightarrow 0.95$ m in increments of 0.025 m.
j) Create a simulation that draws all forces acting on the rod to scale for arbitrary $0.05 \text{ m} < x < 0.95 \text{ m}$. To keep it simple, start by hard coding in a value of x (say $x = 0.50$ m).

To notice:

- Polar form is nice if we want force magnitudes. Perhaps we care about tension in the support cable so we can determine an appropriate size cable for a design project.
- Cartesian form is nice when coding. Perhaps we care about conveying information visually to a co-worker, boss, or someone who hired us for contract work.
- Doing problems algebraically (instead of with one special case of numbers) allows us to quickly test a wide variety of design situations and create visualizations (plots or sims). These visualizations improve our ability communicate effectively with other people.
- With any coding example, it is important to realize the limitations of our model. For example, in this problem we expect the code will produce nonsense if the support cable connects near the left or right end of the rod ($x \approx 0$ or $x \approx L$).

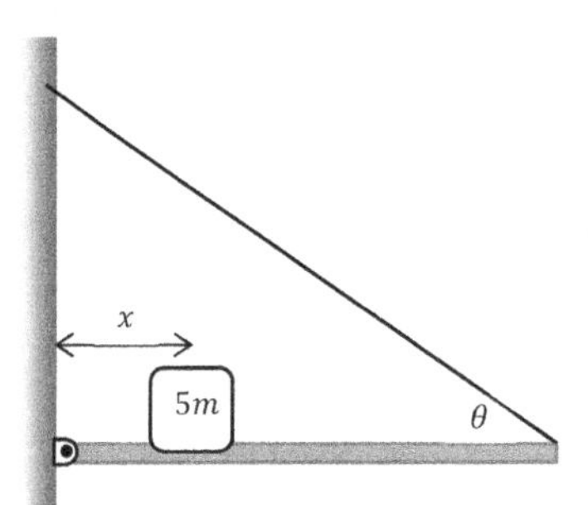

12.2 Now suppose you have a uniform rod with length L and mass m. The left end of the rod is attached to a frictionless pivot point as shown in the figure. The right end of the rod has a string attached to it. This string connects to the same wall as the rod with an angle θ as shown. On top of the rod is an object with mass $5m$. This object's center of mass is distance x from the wall as shown.

a) Draw the FBD for the rod.
b) Why is it ok to use the force $5mg$ acting on the rod when really the bonus object is exerting a *normal* force (not a gravitational force)?
c) Write down the sum of forces for the rod. Also write down the sum of torques for the rod (use the obvious pivot).
d) Suppose $x = L/4$. Determine a relationship between m, θ, and the tension in the string.
e) Suppose the string is rated to support maximum tension of 17.0 N. Assume the angle is 36.9° and we still have $x = L/4$. Assume the bonus object always has exactly 5 times the mass of the rod. What is the greatest mass the bonus object can have without exceeding the string's max tension rating?
f) Assume the support cable is about to snap (assume conditions of part e apply). Determine the force exerted by the pivot on the rod. Answer in Cartesian form.

Going further: Create a plot of tension in the string versus the position of the $5m$ mass.
Going further: Create a simulation showing force arrows on the rod as position of the $5m$ mass goes from $0 \rightarrow L$.

12.3 In each situation shown below a <u>massless</u> rod is supported at one end by a frictionless pivot. In each situation, a block of weight w is placed at various positions on the rod. In addition a light string supports the rod in various ways as shown in each figure. In situations **A** and **D** the block is at the middle of the rod. In cases **B** & **C** the string connects in the middle of the rod.

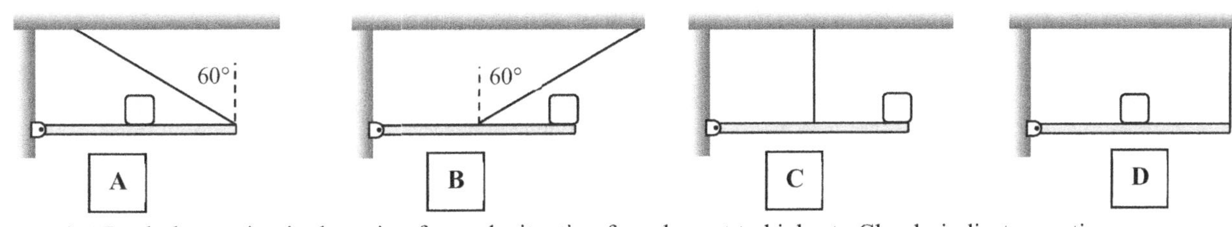

a) Rank the tension in the string for each situation from lowest to highest. Clearly indicate any ties.
b) In which situations, if any, will the reaction force at the pivot have a component pointing left?
c) In which situations, if any, will the reaction force at the pivot have a downwards component? Hint: consider using the center of the rod (or the block…or the string) as a pivot point in each FBD.

12.4 Why should you bend at the knees when picking up heavy objects? You have probably heard the statement, "Don't lift with your back, use your legs." Let's consider the forces on a person's lower back. The principal muscles which support the lower back are the sacrospinal or erector spinae muscles. Very roughly speaking, these muscles act somewhat like a cable under tension connected to the spine. The figure below approximates the human spine and the erector spinae muscles (for more detail do a web search for "erector spinae").

The head/torso/arms are modeled using a rod with mass $m_1 = 2.0$ kg & length $d = 0.92$ m and a dumbbell of mass $m_2 = 5.0$ kg hanging distance d from the pivot. Note: the pivot is the analog of the lumbrosacral joint. Lastly, the cable comes off the rod at 12° distance $x = 0.42$ m from the pivot. This cable models the tension supplied to the spine by the erector spinae muscles. The large mass M attached to the cable over a pulley produces tension in the cable. This part of the demo has no analog in your anatomy.

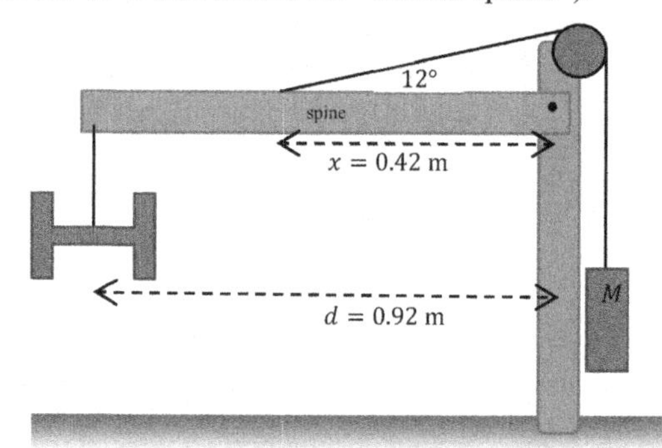

a) Determine the *magnitude* of the torque (about the pivot) produced by the rod and dumbbell. For reference, consider a 150 pound person with average height. The total torque caused by the upper body, arms, and head is probably more than 160 N·m (websearch "erector spinae Harvard").
b) Determine the tension in the erector spinae required to keep the rod balanced in the horizontal position. From this you can compute the appropriate mass M to use in the model.

Think: our model is used for a person bending over without lifting any weight AND the torques are only a third of the typical values for medium sized adults! If a person was not only bending over but also lifting additional weight the tension in the lower back would go up even more!

Last note: what happens if you flex your abs while lifting? It turns out you increase the interabdominal pressure which helps out your lower back (but may cause a hernia if your abs are weak!). Ask a biologist about this...

Disclaimer: in reality, the erector spinae muscles (ESM) are pretty complicated. The ESM are not just one simple cable pulling on the spine. The ESM connect to many different points (about 20) along the spine and ribs. Perhaps this implies, the force in any one bundle is *about* 20 times less than the total force found earlier...but each bundle's connecting angle is slightly different so it is more complicated than described here.

12.5 Ladder Part I Consider a ladder of mass m and length L leaning up against the wall. Assume there is negligible friction between the wall and the ladder. The coefficient of static friction between the ladder and the floor is μ_s. Assume θ is given.

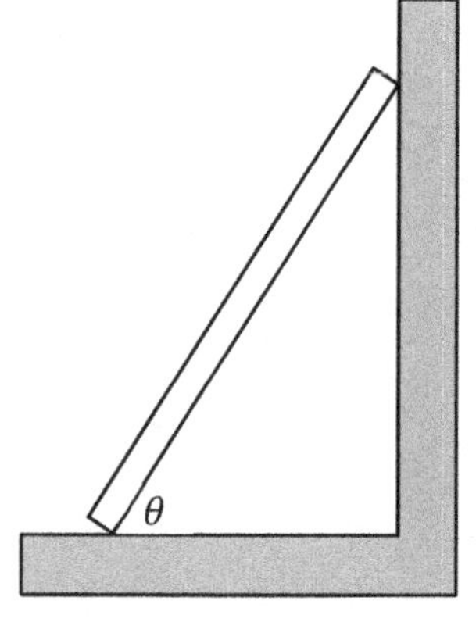

a) Draw a correct FBD of the ladder.
b) Write down sum of forces and sum of torques (use bottom of ladder as pivot).
c) Assuming the ladder is not near the onset of slipping, determine the frictional force acting at the base of the ladder.
d) What is the *minimum* coefficient of friction required to keep the ladder from sliding at this angle? Take note of the variables having no effect on the result.

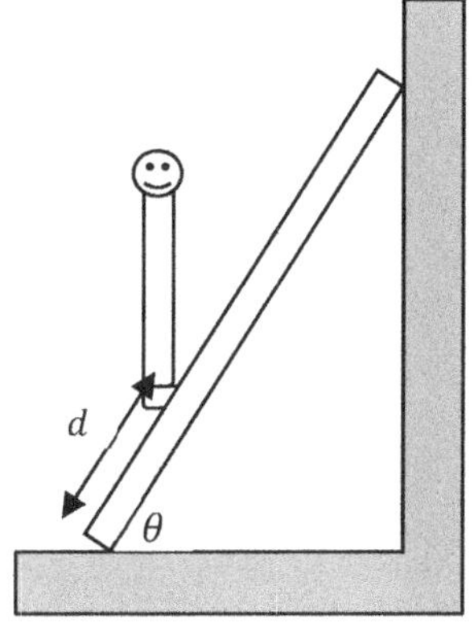

12.6 Ladder Part II Now assume someone, probably a zombie, is climbing the ladder. Assume the zombie has mass m_2 while the ladder has mass m_1. Assume L and θ are unchanged and $\mu_s = \frac{1}{2}\cot\theta$. The zombie is a distance d up the ladder as shown in the figure.

a) Draw a free-body diagram.
b) Write up the appropriate force and torque equations. Use the same pivot as before.
c) How far can the zombie climb the ladder before it begins to slip?
d) Think about ladders in real life. In particular, consider an empty ladder compared to a ladder with a person on the first step. Which is more likely to slide? Does this make sense with your math for this problem?

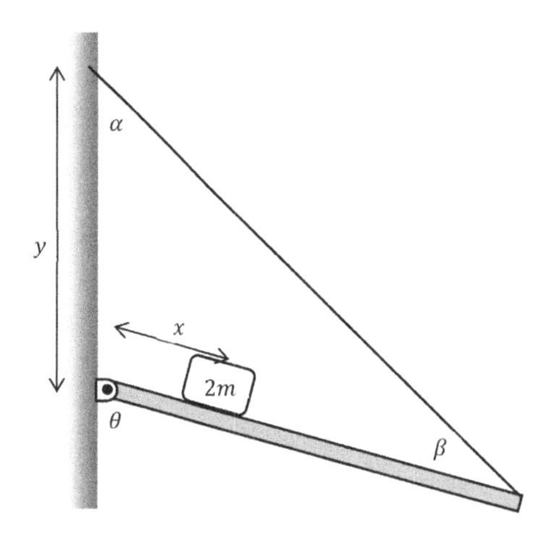

12.7 Draw Bridge Consider a draw bridge of length L and mass m. On the bridge is a box with mass $2m$ distance x from the hinge as shown. The cable connects to the wall a distance of $y = \frac{2}{3}L$ above the hinge. The angle between the bridge and wall is $\theta = 85.0°$ (figure not to scale).

a) Use geometry to prove $\alpha = 52.89°$ & $\beta = 32.11°$.
b) Draw a free-body diagram.
c) Do sum of forces and sum of torques (use hinge as the pivot).
d) Determine the tension in the cable as a function of x.
e) Suppose it is an icy day and the block slides with constant speed off the end. At what point, if any, will the cable snap if it is able to with stand a maximum tension of $4mg$?
f) Assuming the string can still handle a maximum tension of $4mg$, determine the largest block (in terms of mass) that could be placed on the string without snapping the cable for all values of x.

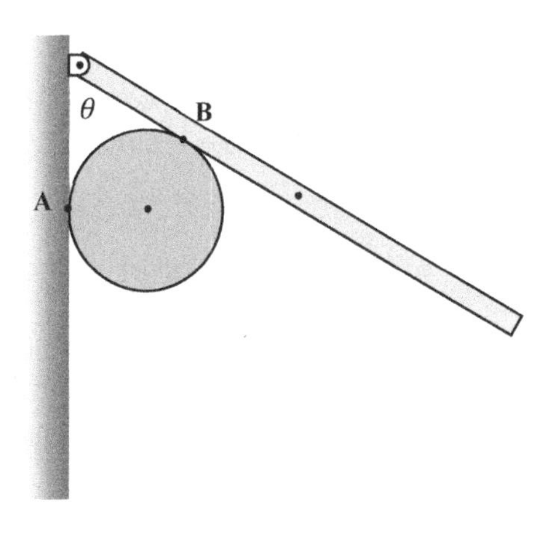

12.8 Class demo: use basketball with lots of grip (not the pink one)
Consider a basketball held in place by a plank that is hinged at the top as shown in the figure. The radius of the ball is R and the length of the rod is $L = 6R$. The points where the ball touches the wall and plank are called points **A** and **B** respectively. Point **B** is ¼L from the end of the plank. The mass of the ball is m_1 while the mass of the plank is m_2. You may assume the hinge is frictionless.

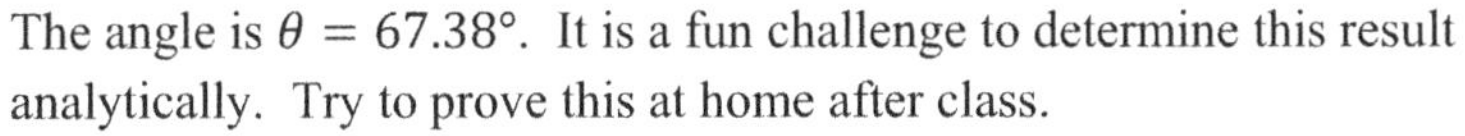

The angle is $\theta = 67.38°$. It is a fun challenge to determine this result analytically. Try to prove this at home after class.

a) Sketch an FBD of the forces on the ball. Do not assume the ball is about to slip.
b) Sketch an FBD of the forces on the plank.
c) Determine the sum of torques on the ball using the center of the ball as a pivot.
d) Determine the sum of forces on the ball.
e) Determine the sum of torques on the plank using the hinge as the pivot.
f) Determine the sum of forces on the plank.
g) When a ladder leaning against vertical wall is on the verge of slipping we assume the frictional forces at the floor and at the wall are both given by $f = \mu n$. I don't believe the same can be said about this problem. What do you think? Can you imagine a situation where $f = \mu n$ at point **A** but not point **B** (or vice versa)? Is my intuition incorrect? What are your thoughts?

12.9 A rod of mass m and length L is supported by a cable and frictionless pivot as shown. A cable of length $\frac{2}{3}L$ supports the rod. The cable comes off the rod at a right angle as shown. The cable connects to the rod distance $\frac{3}{10}L$ from the pivot.

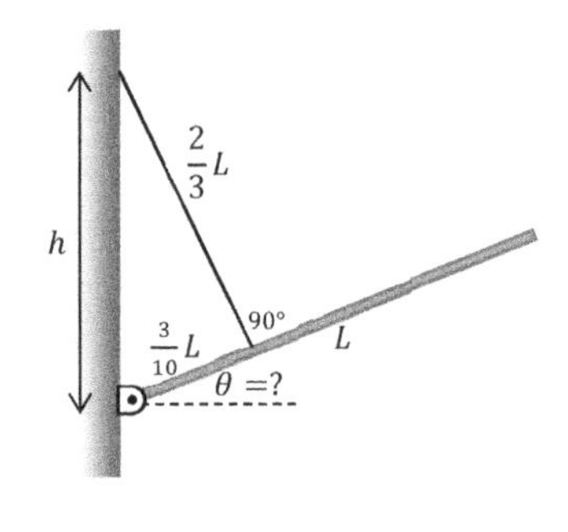

a) Determine the distance h shown in the figure. Answer as a simplified fraction times L.
b) Determine θ, $\sin\theta$, & $\cos\theta$.
c) Use sum of torques about pivot to determine tension in the string.
d) Determine force equations using a standard xy-coordinate system (x is horizontal…not along incline).
e) Determine equations for the reaction force components at the pivot. Use your results to determine the magnitude and direction of the reaction force.

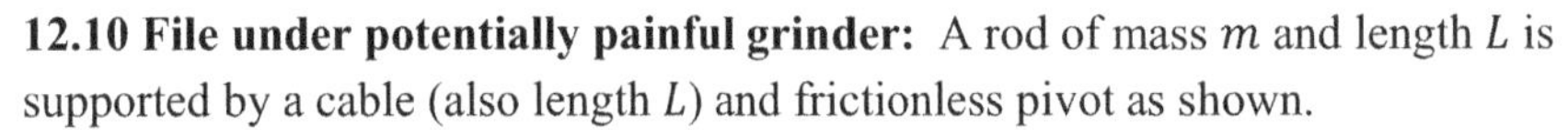

12.10 File under potentially painful grinder: A rod of mass m and length L is supported by a cable (also length L) and frictionless pivot as shown.

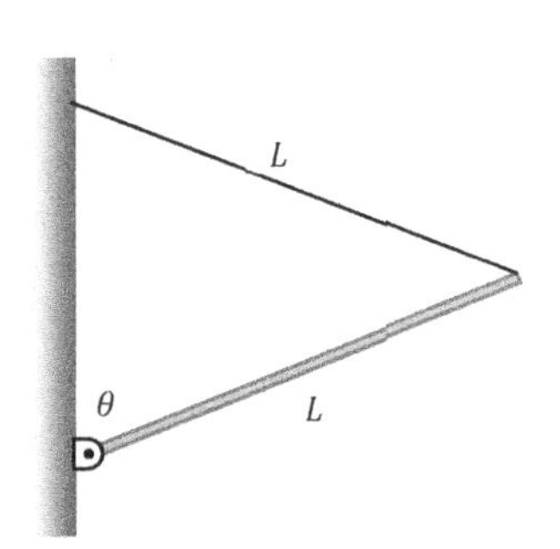

a) Determine torques about pivot.
b) Determine force equations using a standard xy-coordinate system (x is horizontal…not along incline).
c) Determine the tension in the string.
d) Determine the reaction force components at the pivot. Use your results to determine the magnitude and direction of the reaction force.

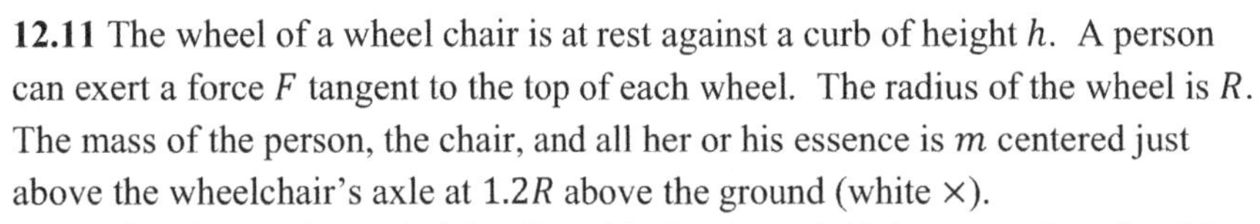

12.11 The wheel of a wheel chair is at rest against a curb of height h. A person can exert a force F tangent to the top of each wheel. The radius of the wheel is R. The mass of the person, the chair, and all her or his essence is m centered just above the wheelchair's axle at $1.2R$ above the ground (white ×).
Determine the maximum height allowable for the curb if the person is to be able to roll the wheelchair over the curb.
Hint: if the wheel chair *just barely* lifts off, think of it as rotating about what point with $\alpha \approx 0$? Since $\alpha \approx 0$, statics still *approximately* applies!

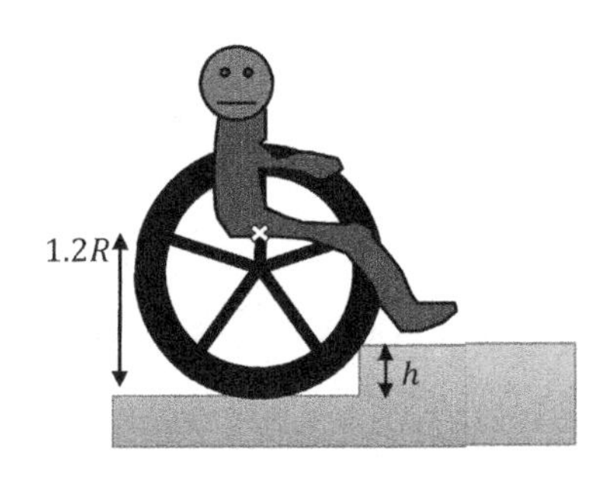

12.12 A nearly empty spool of wire with mass M is in the corner of a room as shown in the figure at right. The outer radius of the spool is R while the inner radius is $R/3$. There is negligible friction between the wall and the spool while the coefficients of friction between the spool and the floor are μ_s and μ_k. The full moon goes behind some clouds just long enough for a werewolf to pull on the end of the wire coming from the spool at angle ϕ from the vertical as shown. Note: since the spool is nearly empty the wire can be considered as attached to the inner radius of the spool. Furthermore, you may assume the center of mass of the spool is at its geometric center. Figure not to scale.

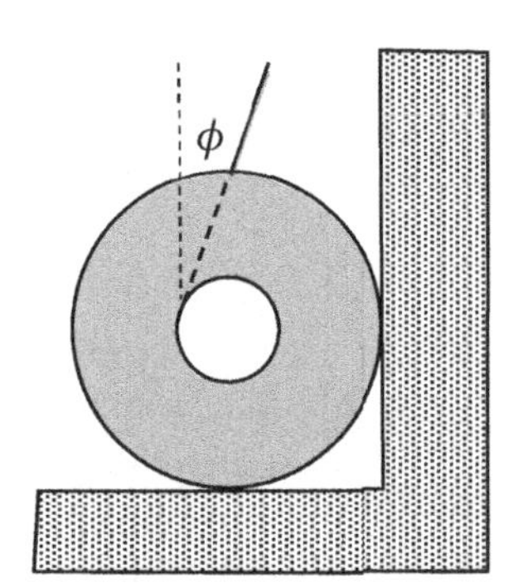

a) With what *minimum* tension must the werewolf pull on the wire to cause the spool to spin? Answer in terms of the known variables and g. You may assume it spins and doesn't lift off the table.
b) **Challenge:** Will the spool ever tend to lift-off instead of spin? Does it depend on the μ_s and/or μ_k?
c) **Challenge:** Is your previous work valid for all values of ϕ (including both positive and negative angles)? Does the range of viable angles depend on the values of μ_s and/or μ_k? **Explain**.

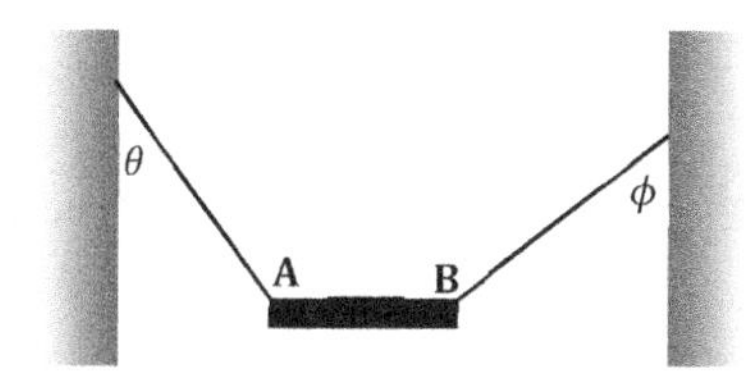

12.13 A non-uniform, horizontal beam of is supported by two vertical strings. The position of the center of mass of the beam is unknown. We do know the left string makes angle $\theta = 36.87°$ while the right string makes angle $\phi = 53.13°$. The beam has mass m and length L. For ease of communication, call the left end of the beam point **A** and the right end point **B**.

a) Which tension is larger? Are they the same? Circle the best answer. How can you tell without doing the entire problem?

Left string	Right string	Same tension	Impossible to determine

b) Is the center of mass of the rod closer to the right or left end of the rod? How can you tell without doing the entire problem?

Closer to right end	Closer to left end	Exactly in the middle	Impossible to determine

c) Determine the tension in each string. Express each answer as a four sig fig number times mg.

d) Determine the distance from the left end of the beam to the center of mass of the beam. Answer as a four sig fig number times L.

12.14 Suppose a thick steel rod is to be mounted to a wall such that it will extend horizontally. The rod will attached to the wall using a frictionless pivot point. The rod can be purchased in various lengths and has linear mass density of $250\frac{\text{kg}}{\text{m}}$. A winch, located distance $x = 7.071$ m above the pivot, attaches a cable to the end of the rod. For now, assume the rod has length L. The cable has maximum tension rating 15.0 kN.

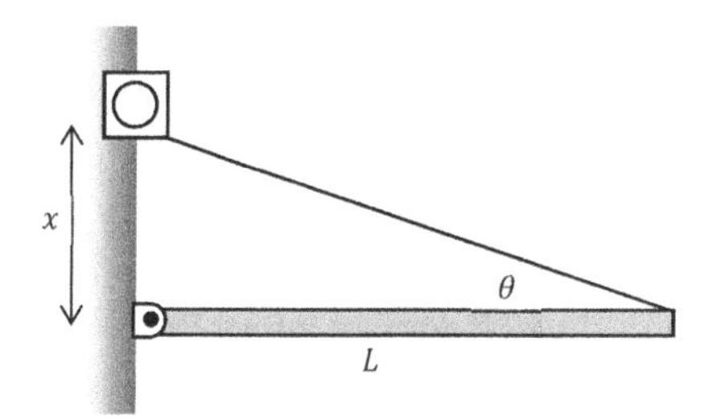

Note: the solution is written up algebraically. Only the final answers have numbers. Do your work algebraically as well so you can follow along.

a) Determine expressions for $\sin\theta$ and $\cos\theta$ in terms of x and L.

b) Draw the FBD of the rod. Determine a torque equation about the pivot as well as the force equations in both the x- and y-directions. Do not use θ in your equations.

c) Determine the longest rod that can be suspended.

d) Determine the reaction forces at the pivot.

Think conceptually: For longer rods, obviously there will be more mass to support. Furthermore, for longer rods the cable will come off the rod at smaller θ. This means the component of tension perpendicular to the rod decreases while the required tension is increasing. Both factors increase the likelihood the cable will break. I suspect the tension in the cable will grow rapidly with the length of the rod.

It would be neat to plot T *vs* L for this scenario.

Alternatively, create a simulation wherein force arrows acting on the rod are drawn as L gradually increases.

12.15 Ladder Part III Consider a ladder of mass m and length L leaning up against the wall. The coefficient of static friction between the ladder and the floor is μ_s. The same coefficient now exists between the ladder and the wall! Determine the minimum coefficient of friction required to prevent the ladder from sliding. Assume θ is given.

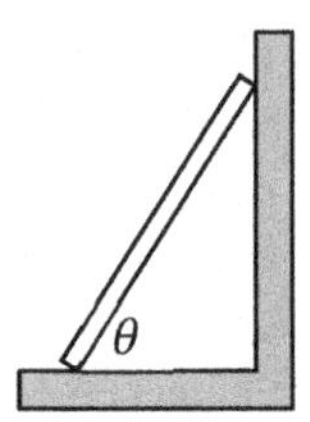

12.16 The one time we actually use fictitious forces

Consider a car of mass m rounding a curve on a flat track. The center of mass of the car is a distance y above the ground. The distance between the outer edges of the wheel's (essentially the car's width) is w. The car is travelling at speed v. Assume the coefficients of sliding friction between the wheels and the road are both μ.

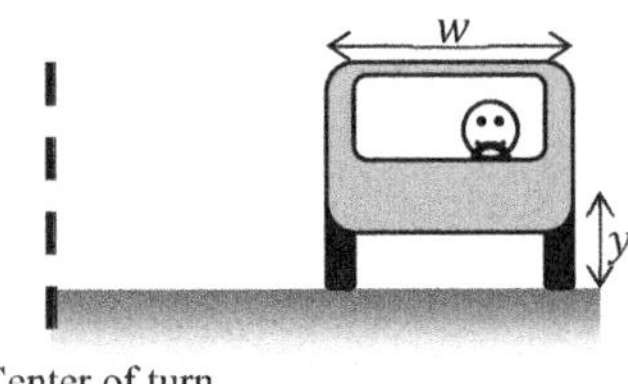

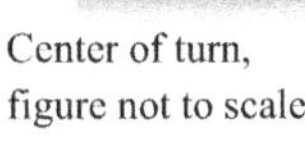

Center of turn, figure not to scale

a) What is the smallest turn it can negotiate without <u>sliding</u>?

b) What is the smallest radius turn the car can travel at this speed without tipping over? Hint: to evaluate this problem it helps to eliminate centripetal acceleration by creating a fictitious force. If one uses the fictitious force, the tipping problem can be considered with a statics problem! Hint^2: at the onset of tipping, only one set of wheels is making significant contact with the ground…which set? Hint^3: you could consider the center as a pivot as well!

c) Which of the above answers are affected by mass?

d) Do the dimensions of the car significantly affect the speed at which *sliding* onsets?

e) Do the dimensions of the car significantly affect the speed at which *tipping* onsets?

12.17 Zombie Ladder Revisited

Now consider a zombie of mass m standing on a ladder a distance d from the bottom. Assume the mass of the ladder is negligible compared to the zombie. The feet of the ladder are width w apart. The top of the ladder is h above the ground. At $\frac{2}{3}h$, a tie rod is used to keep the ladder from falling down. Assume friction between the floor and the ladder is negligible. Figure not to scale.

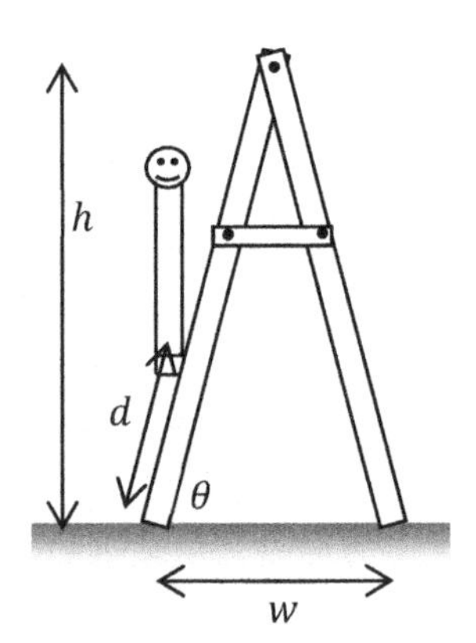

a) Determine $\tan\theta$ (not θ) in terms of w and h. For the rest of the problem you may assume θ is given since it is easily determined by this result.

b) Draw FBDs for each half of the ladder.

c) Do sum of forces for the left half of the ladder.

d) Do sum of forces for the right half of the ladder.

e) Do sum of torques at the top for the left half of the ladder.

f) Do sum of torques at the bottom for the right half of the ladder.

g) What is the force in the tie rod as a function of d? Doe the tension in the tie rod increase or decrease as the zombie climbs the ladder.

h) What is the normal force on each half of the ladder as a function of d?

i) Do your answers to the previous parts make sense at the extreme limits of $d = 0$ and/or $d = L$?

12.18 A sphere of mass m and radius R is supported by a cable on an incline. Assume friction between the incline and the sphere is negligible. The incline angle is α above the horizontal while the cable is angled at angle β from the incline as shown. The cable connects to ball such that the cable extends radially away from the center of the ball. To be clear, the lengths of the cable and incline are unknown.

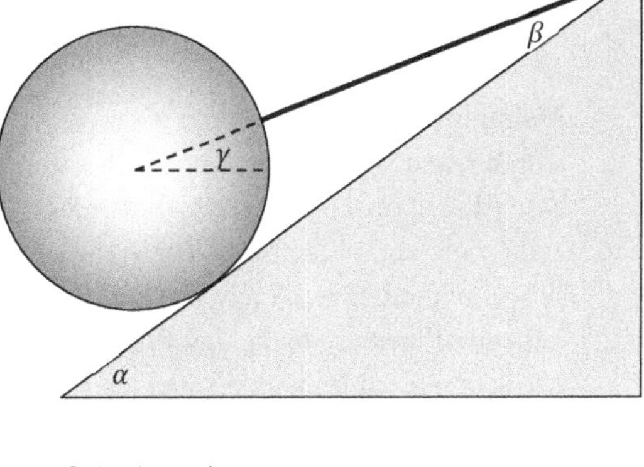

a) Determine the angle γ indicated in the figure in terms of the angles α & β.

b) Write a correct sum of torques equation about the point of contact between the ball and the incline.
Note: writing $\Sigma\tau = 0$ is not worth any points.
Note: for this and all subsequent parts you may assume all angles (α, β & γ) are known.

c) Write a correct sum of forces on the ball in the *horizontal* direction using the standard $\hat{\imath}$ direction.

d) Write a correct sum of forces on the ball in the *vertical* direction using straight up as the vertical direction.

e) Determine the tension in the cable in terms of m, g & any of the angles you need.

WATCH OUT! This is an unusual Chapter 12 problem in that one can solve part e without doing sum of torques. It is foolish to expect a Chapter 12 exam question which doesn't involve torque.

12.19 TIPPING OVER A cubical crate with sides 1.50 m is at rest on the incline shown. The center of mass of the crate is 0.300 m above the crate's geometrical center (black dot in figure). The angle θ is gradually increased until the block either tips or slides.

At the onset of tipping:

a) Where is the normal force acting on the block?

b) What must be true about the line of action for the object's weight?

c) Why is it reasonable to consider an equilibrium problem at the onset of tipping?

d) At what angle would the block tip? Assume sufficient friction is present such that block tips instead of slips.

e) Now assume you have three different boxes with uniform mass distribution. One of them is twice as tall as it is wide, one is cubical, and one is twice as wide as it is tall. Each one is separately placed on an incline where the incline angle is gradually increased as in the previous problem. What minimum coefficient of friction is required for each of the boxes to tip over instead of slide? Which type or types of boxes will almost always slide and which will be most likely to tip over?

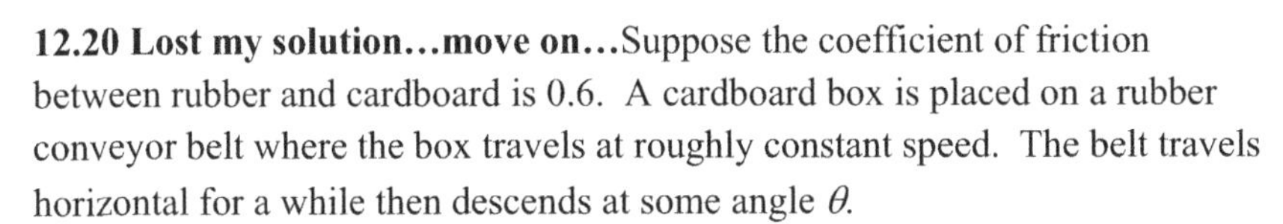

12.20 Lost my solution...move on... Suppose the coefficient of friction between rubber and cardboard is 0.6. A cardboard box is placed on a rubber conveyor belt where the box travels at roughly constant speed. The belt travels horizontal for a while then descends at some angle θ.

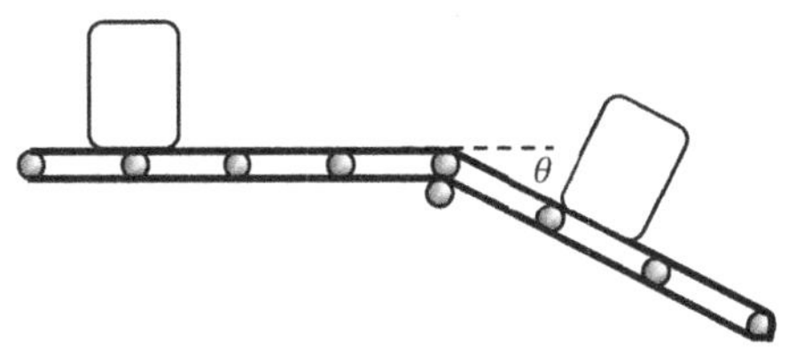

a) What is the largest angle that can be used before the boxes will slide on the conveyor?

b) Assuming uniform packages, what is the maximum ratio of height to width to prevent the boxes from tipping on the angled portions of the conveyor?

c) Suppose the belt is started from rest. Suppose that the boxes have the maximum height to width ratio from the previous problem. What is the maximum acceleration the belt can have to prevent boxes from tipping on the flat section of the track?

d) Now consider an identical box with identical acceleration that is located on the incline? Will it tip over or not? Will it slide? If it doesn't slide or tip, what is the magnitude of the friction force acting on the box?

12.21 Disclaimer: I don't work with cranes and have zero expertise on this matter. That said, we can try to scratch the surface even though I know this design is impractical and oversimplified.

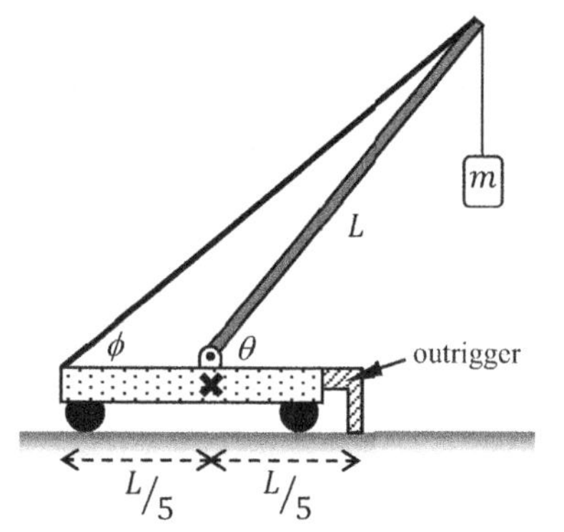

A simplified model of a crane is shown. The boom of the crane (dark grey rod) has length L and is angled θ above the horizontal. The boom attaches to the center of the base (dotted rectangle) with a frictionless pivot. The crane is supported by an outrigger at the right end of the base. The crane supports a load with mass m with a winch not shown in the figure. A guy line (thick black line) connects a winch (not shown) at the left end of the base to the tip of the boom. Assume the guy line and the load line are two separate cables with distinct tensions. The mass M of the base includes the cabin, engine, and two adjustable counter-masses. The center of mass of all that crap is located at the black **x** in the figure just below the boom's pivot. The mass of the boom and guy line are negligible. Note: figure not to scale.

a) When the crane picks up objects from the ground they must lie to the right of the outrigger. Determine the largest angle θ the crane can use for objects on the ground.
b) Determine a relationship between θ and ϕ. Hint: use geometry and/or trig.
c) Suppose this is a desktop demonstration crane. The guy line is rated to withstand maximum tension $T_{max} = 20$ N. The crane is using angle $\theta = 40°$. Determine the mass M required to keep the crane from tipping over with the wheels barely touching the ground. Hint: consider not only an FBD for the boom but also an FBD for the entire crane. Try to do the problem algebraically as long as possible and stick in numbers at the end. This may help answer the questions below.
d) What friction force is required at the point of contact between the outrigger and the ground to keep the crane in equilibrium?
e) What happens to the crane if the mass M is above or below the value determined in part c?
f) How does changing the angle θ affect the tension in the guy line? Explain why.
g) Suppose m is not static but instead moving, how is the tension in the guy line affected? Explain all cases.
h) **Challenge:** What are the pros and cons of using a slightly longer outrigger?
i) **Challenge:** What are the pros and cons of using a shorter boom?
j) **Challenge:** Consider the two different crane designs shown at right. The one on the right uses a hydraulic piston to raise and lower the boom as opposed to using the guy line. Will the guy line or the hydraulic piston be exposed to greater forces? What are the pros and cons of the two designs? Defend your answer.

12.22 A square block of side L and mass m has tension applied at $\frac{3}{4}L$ above the ground. Show the torque equation and forces equations give

$$mgx - \frac{mgL}{2} + \frac{3}{4}LT = 0$$

From this one finds

$$x = \frac{L}{2} - \frac{3LT}{4mg}$$

Notice this makes sense in the limit $T \to 0$.

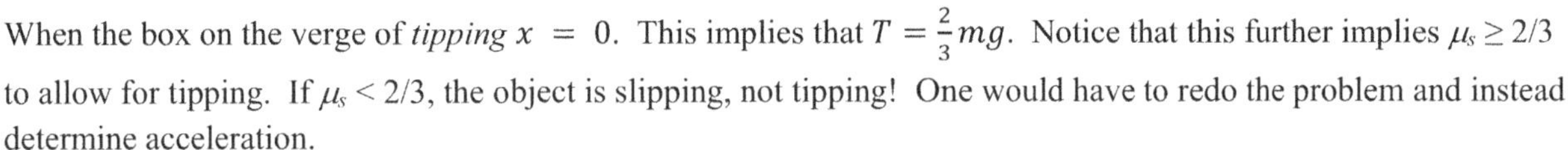

When the box on the verge of *tipping* $x = 0$. This implies that $T = \frac{2}{3}mg$. Notice that this further implies $\mu_s \geq 2/3$ to allow for tipping. If $\mu_s < 2/3$, the object is slipping, not tipping! One would have to redo the problem and instead determine acceleration.

Cool trick: Consider the tension applied to the upper left corner of the block of width w and height h. Suppose we know both the magnitude and angle for the tension force. First split the tension force into components and then use lever arms on each component separately! This trick can sometimes simplify some nasty calculations.

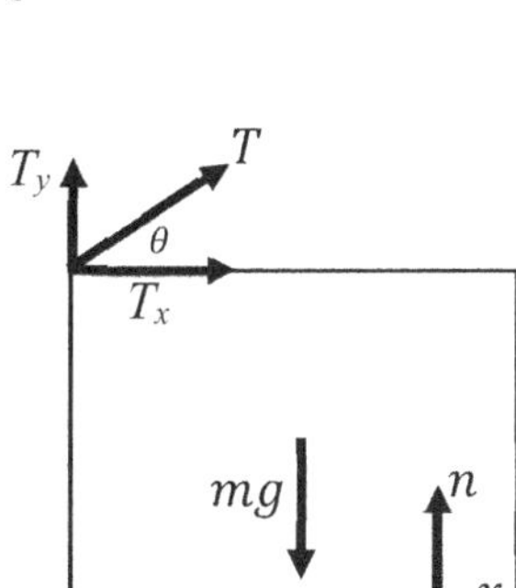

12.23 Not solved yet but fun to think about Yo-yo placed in *right-angle* corner (left figure). Coefficient of static friction is μ between wall and spool and between floor and spool. Inner radius (were string attaches to spool) is r while outer radius (part of spool in contact with floor and wall) is R. What is largest force magnitude F applied to string for which yo-yo won't spin. Watch out: check when finished if $F > mg$. If $F > mg$ then yo-yo would lift-off ground making the problem more complicated than simple slipping!!!!

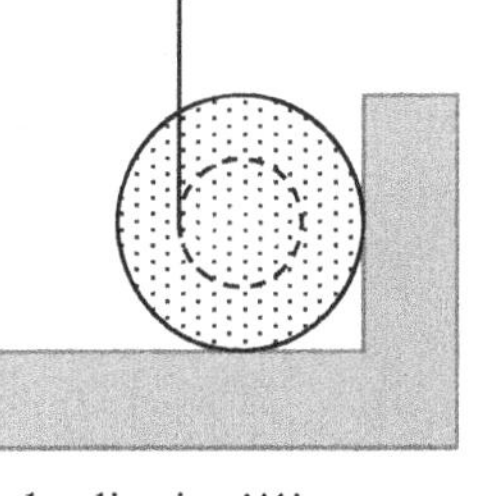

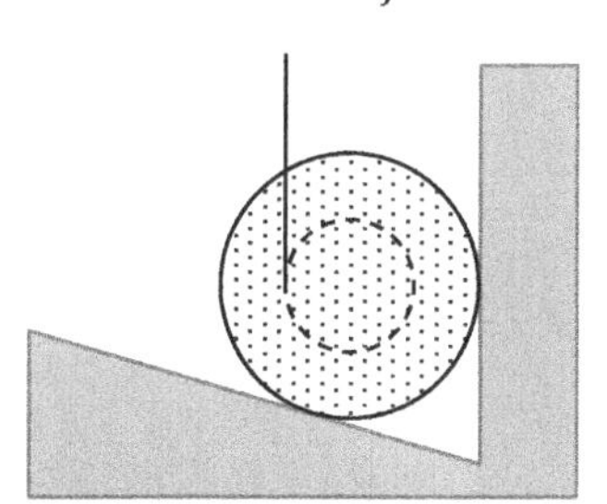

Repeat if corner has *acute* angle (right figure). Take a guess…will the yo-yo be more or less likely to slip? Now figure out.

12.24 Not solved yet but fun to think about In each picture below the woman has a weight of 200 lbs while the ladder has a weight of 50 lbs. Let us assume that the ladder is <u>not</u> on the verge of slipping and there is negligible friction between the wall and the ladder. The woman is 75% of the distance up the ladder. In each case, draw the FBD of the ladder and determine the <u>**magnitude**</u> and <u>**direction**</u> of force exerted by the ground on the ladder. Also determine the force exerted by the wall on the ladder for each case.

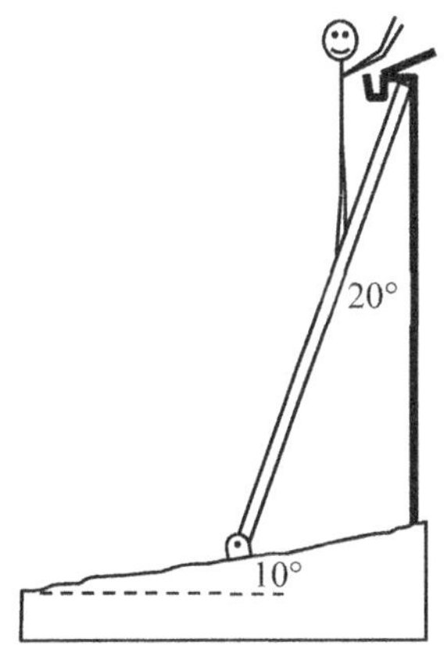

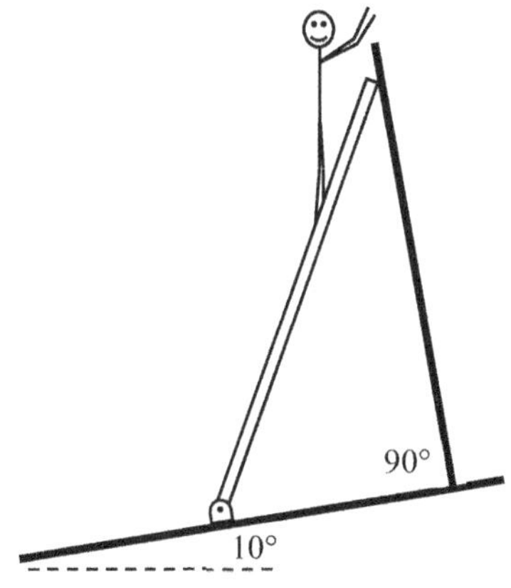

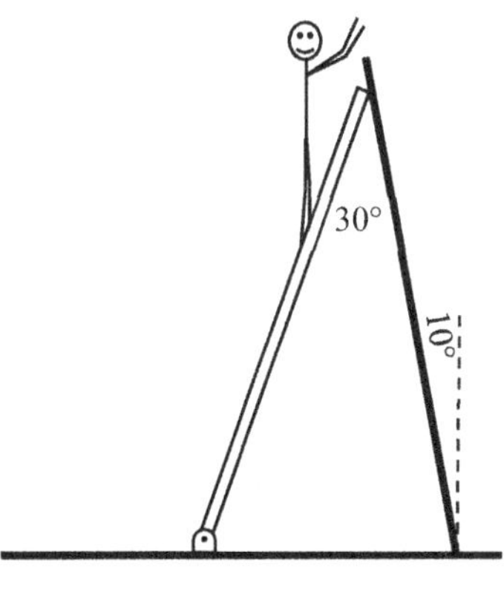

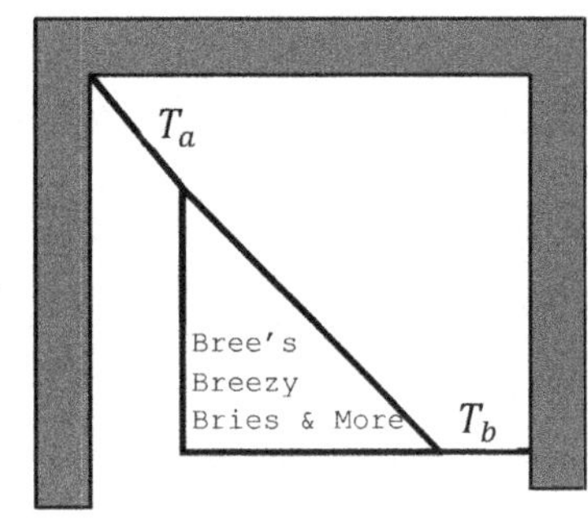

IMPOSSIBLE 1? Your friend Bree has a cheese shop and wants to hang a 22-pound ($mg \approx 100$ N) sign. The sign is to be made of a uniform, triangular plate that looks like a piece of cheese. The sign is a 45-45-90 triangle with a short sides of $L = 1.00$ m.

Bree knows that the forces on the sign must balance both left and right as well as up and down. She also recalls that the center of mass of a right triangle is 1/3 of the distance from the fat end (both vertically and horizontally). The cheese business was really taking off so she dropped this class before getting to torques & static equilibrium and now you have to get involved.

Bree tells you she wants to hang the sign using light cables as shown in the figure. She determines the tensions in each cable must be $T_a \approx 140$ N & $T_b \approx 100$ N. Her cousin Camembert says she messed up T_a by a factor of 3. The building owner, Penamellera, overhears the debate and thinks this situation could never work. What's going on here? Who is right? Is there more than one correct answer, only one, or is this impossible?

Sum of forces <u>horizontally</u> gives

$$T_a \frac{\sqrt{2}}{2} = T_b$$

Sum of forces <u>vertically</u> gives

$$T_a \frac{\sqrt{2}}{2} = mg \approx 100 \text{ N}$$

Using these two results together gives Bree's results.

Doing sum of torques about the bottom *left* corner of the sign we find

$$LT_a \frac{\sqrt{2}}{2} - \frac{L}{3} mg = 0$$

$$T_a \frac{3\sqrt{2}}{2} = mg$$

This is where Camembert is coming from. Notice these results contradict Bree's results!

Finally, doing sum of torques about the bottom *right* corner of the sign we find only one torque! Notice the lines of action for both tensions run through the bottom right corner. It is impossible for the torques to be balanced about the bottom right corner if there is only one torque!

Clearly this is an impossible arrangement. It is impossible to simultaneously balance forces and torques using this arrangement.

IMPOSSIBLE 2? Now consider a 100 N sign shaped like a 30-60-90 triangle (uniform mass distribution). Assume the longest side is L. Is it possible to hang it in the orientation shown? If so, what are the tensions in each cable? If not, explain why it cannot be hung in such a manner.
As a challenge, prove the x-coordinate of triangle's center of mass is distance $\frac{5}{12}L$ from the upper left corner. The center of mass is indicated by the black dot in the figure.

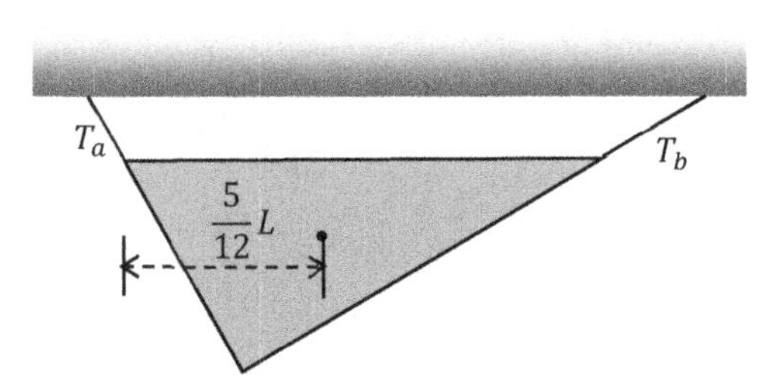

Answer: Doing torques about the upper *left* corner gives $T_b = 83.33$ N. Doing torques about the upper *right* corner gives $T_a = 67.34$ N. Using these values of forces I checked to see if the horizontal forces balanced. They did not. The sign cannot be hung in this manner. Using these angles, you cannot balance forces and torques simultaneously. Therefore the system <u>cannot</u> be in static equilibrium using the arrangement shown.
Note: torques about the bottom corner gives only one torque (check lines of action).

Again, this is a physically impossible scenario!!!

Overdetermined and underdetermined systems

For our physics problems we *typically* have three equations for static equilibrium (forces in the x- and y-directions and torques about a point). We could get a fourth equation by choosing a different point and redoing the torque equations. *Usually*, these fourth equations are *not* linearly independent. That means you can combine three of the equations in some manner to create the fourth one. This means having that fourth equation is no better than our original three.

If we have only three linearly independent equations to work with, we can only solve systems with three unknowns. If we have more than three unknowns the system is said to be *underdetermined* and there is no way to get a solution. If the system has *fewer* than three unknowns, the system is *overdetermined* and *may* give contradictory results.

Notice that both problems on the previous page were *overdetermined* systems. Consider, however, if each of the previous problems had one cable with an *unknown* angle. Each system would then have *three* unknowns, *three* linearly independent equations, and *would* be solvable.

As an exmple, consider the sign shaped like a 30-60-90 triangle. The mass of the sign and the two tensions were unknown but the angles were given. This problem is *overdetermined.* We end up with more linearly independent equations than variables. A way to see this is to consider torques about point **C** in the figure at right (equation (1) below). Then do forces in the x- and y-directions (equations (2) and (3) below respectively). We find

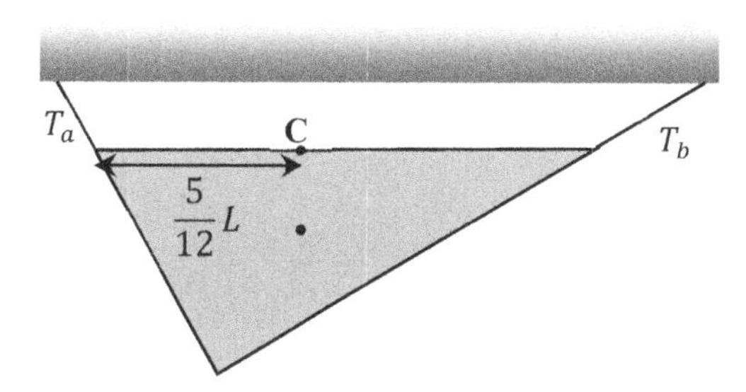

$$\frac{5}{12}LT_a \sin 60 = \frac{7}{12}LT_b \sin 30 \qquad (1)$$

$$T_a \cos 60 = T_b \cos 30 \qquad (2)$$

$$T_a \sin 60 + T_b \sin 30 = mg \qquad (3)$$

Notice the first two equations give inconsistent results for T_a and T_b regardless of the unknown m.

Finally, consider adding in a fourth equation using torques about the upper left corner of the triangle. We find

$$\frac{5}{12}Lmg = LT_b \sin 30 \qquad (4)$$

Notice that by subbing in equation (3) for mg in equation (4) you get back equation (1). This shows equation (4) is not linearly independent from the previous three equations.

What would happen in real life if you tried to hang a triangular plate like the one shown above?
You would never get it to hang exactly like the figure. At least one of the strings would necessarily form a different angle. Otherwise the system can't have balanced forces and balanced torques simultaneously!!!

Going further: to fix this problem can you simply let both strings be vertical? If yes, verify. If not, at what sets of angles will this problem work? Is there more than one option? I have no clue but I bet you could figure it out.

Note: a truss system with several rods (instead of a single rod) will have more than three linearly independent equations. An example of such a system was the ladder with the tie rod.

12.25 Now consider a door hanging with two attachment points. The top attachment point is a hinge that exerts reaction forces in both the x- and y-directions. The lower attachment point is a roller. It will only exert a force perpendicular to the wall. If there was a vertical force on the roller, it would simply roll to a new position and eventually reach equilibrium with no vertical force. The upper hinge is at point **A** distance $L/4$ from the top while the roller is at point **B** distance $L/4$ from the bottom. The weight of the door is mg. The width of the door is 1/3 of the height L.

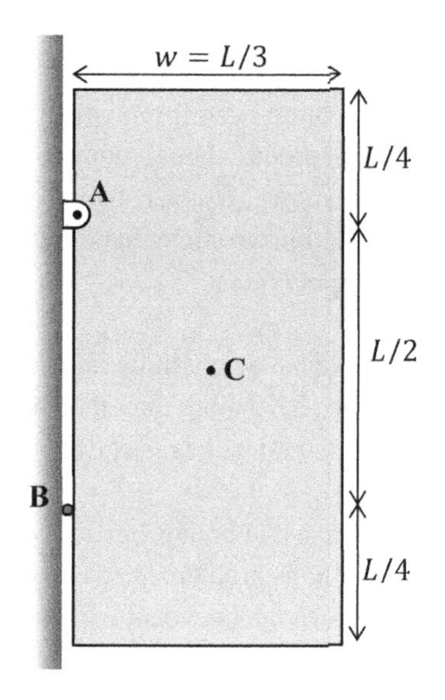

a) Determine reaction forces A_x, A_y, and B_x. Answer as fractions times mg.

b) In what direction does the *hinge* exert a force on the door?

c) How would your solution change if, instead of a *roller*, the bottom point used a *hinge*?

STRESS & STRAIN

Solids have a definite shape (a block of wood)

Crystalline solid (grain of salt, highly ordered structure) vs amorphous solid (glass, random structure)

Liquids will take on the shape of a container (bottle of liquid)

Gas & plasma are free-form with no definite form or shape.

Gases and liquids are both considered fluids for this class since they both flow in the same way.

Plasma is most common form of matter as it is found around stars.

Dark matter is perhaps 25% of the erg in the universe.

Dark energy is possibly 70% of all matter in the universe.

Solids are often modeled as a bunch of tiny particles connected by springs.

Show matter model demo & simulation online.

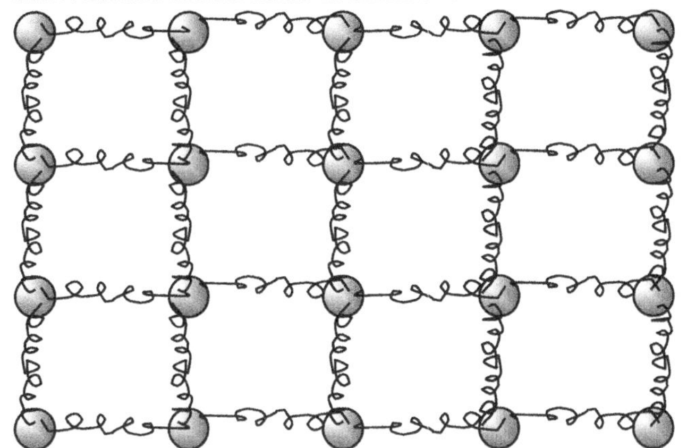

The spheres represent atoms. The springs represent the electrical forces between the atoms.

If you were to try and stretch this group of atoms…would it be easier to stretch it vertically or horizontally?

This model is accurate to some degree because when you try to stretch a solid it will actually stretch! The amount is stretch is often so small your eye can't see it. There is an equation that governs this amount of stretch just like a spring! For a spring $F = kx$. Think: if longer, if larger cross-section, diff materials…how would these affect a spring constant k?

For a solid

$$F = \frac{AE}{L_0}\Delta L$$

where the effective spring constant is $k = \frac{AE}{L_0}$. Here $x = \Delta L$ is the amount of stretch, A is cross-sectional area of wire, L_0 is unstretched length, and E is called the modulus (in this case Young's modulus). Notice, according to this equation, it is easier to stretch long, skinny (small cross-sectional area) objects like wires.

All this is swell, but it is usually written another way:

$$\frac{F}{A} = E\frac{\Delta L}{L_0}$$

Often is discussed in yet another way:

$$stress = \frac{F}{A}$$

$$strain = \frac{\Delta L}{L_0}$$

$$\frac{stress}{strain} = modulus$$

Stress is a force per unit area causing a deformation, strain is a measure of the amount of deformation

Note: It is common to express strain as a % (e.g $\frac{\Delta L}{L_0} = 0.2\% = 0.002$).

Note: Some resources might use Y instead of E for Young's modulus.

12.26 A 10 cm long piece of taffy will stretch about 0.6 cm with 20 N of force (imagine clamping a 4 lb weight to some taffy). The taffy is about 5 mm thick and 2 cm wide.

a) Determine Young's modulus for the taffy. Keep track of the units!
Is this a big number? To get a feeling for the size of these numbers, the Young's modulus of one type of concrete is $E = 20$ GPa where 1 Pa $= 1\frac{\text{N}}{\text{m}^2}$.

b) Would it be easier, harder, or the same difficulty to stretch the taffy parallel to the width (the 2 cm dimension)? Explain why.

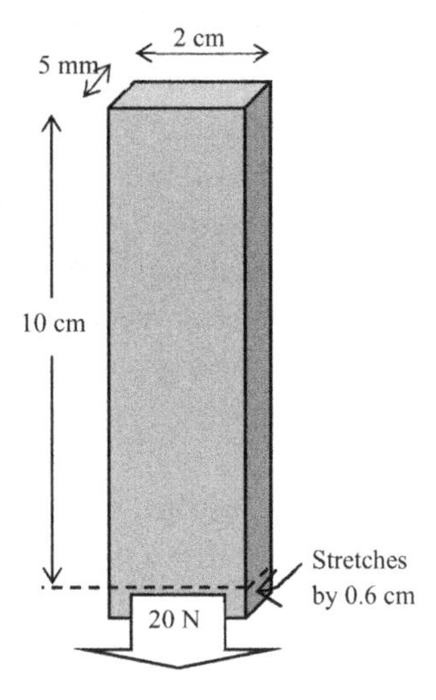

12.27 A Schedule 40 PVC pipe has outer diameter D, inner diameter d, and height h.

a) Determine an expression for the surface area of the end. The end is shaded with diagonal lines in the figure.

b) A mass m is placed on top of the pipe. Assume the mass of the pipe itself is negligible compared to m. Determine an algebraic expression for the resulting compression distance of the pipe.

c) For 1-inch nominal pipe one website stated $D = 1.315$ in & $d = 1.029$ in. Nominal means size in name only; a 1-inch nominal pipe is approximately but not exactly 1 inch in diameter. Assume $m = 20.0$ kg and $h = 24.0$ in and $E = 4 \times 10^5$ PSI. Determine the compression distance.

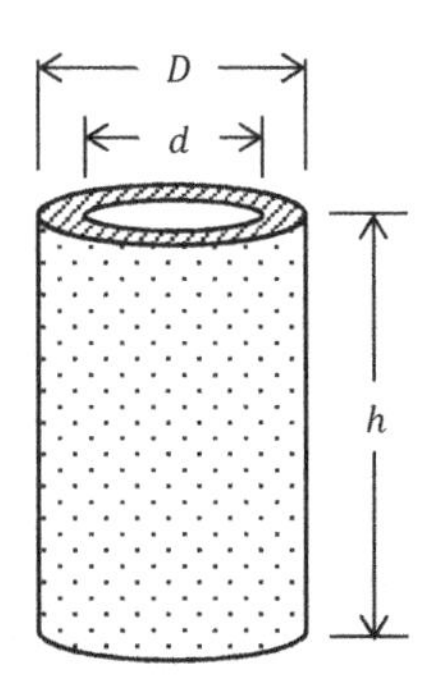

12.28 Consider rods **1** and **2** shown at right. **Assume the left end of each rod is clamped in place to keep the rod motionless.** Tension force of magnitude F_1 is applied to the right end of **1** while tension force of magnitude F_2 is applied to **2**.

a) Determine the ratio of F_1 to F_2 if both rods are stretched $0.1L$.

b) Determine the ratio of F_1 to F_2 if both rods are stretched 5%.

c) In which of the above cases do the rods experience the same strain?

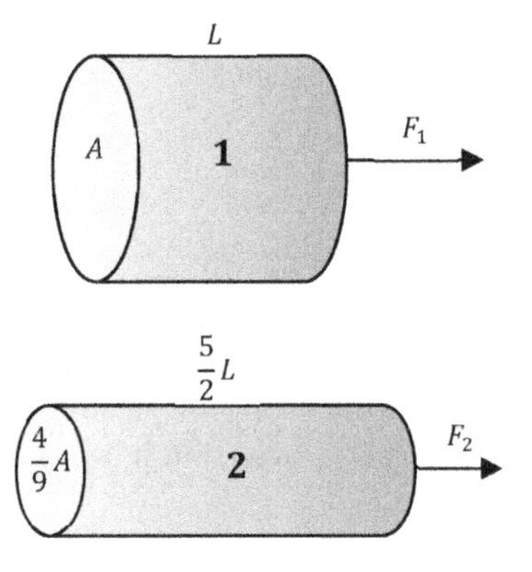

Similar results apply in your book for SHEAR strain. Think about a book with one force along the top and opposite force along the bottom causing the book to take on a slanted shape.

Bulk modulus: similar to Young's modulus but can also be applied to fluids.

Summary: when an object is put under stress it experiences strain.

Tensile stress(1D stretching and compressing) use Young's modulus: $\frac{F}{A} = E\frac{\Delta L}{L_0}$

Shear stress(bending) use shear modulus: $\frac{F}{A} = S\frac{\Delta x}{h}$

Compressing volumes use bulk modulus: $\Delta P = -B\frac{\Delta V}{V_0}$ here ΔP is change in pressure which has same units as $\frac{F}{A}$

Aren't you curious about the minus sign? Does it make sense?

This topic is covered much more in depth in a Strength of Materials class.

12.29 A piece of elastic cord has initial unstretched length L. A mass m is attached to the end of the cord and the system is allowed to reach equilibrium. At equilibrium the cord is stretched distance x and has diameter D.

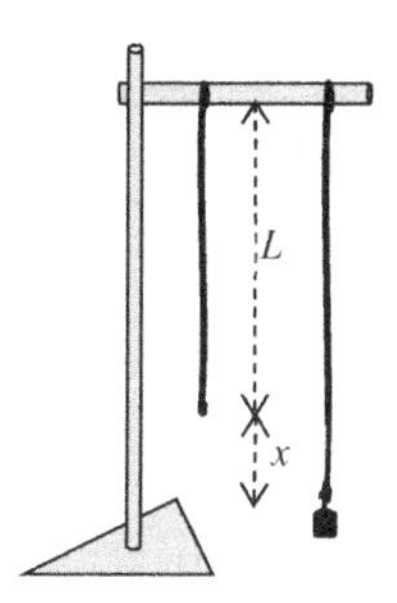

a) Determine the spring constant of the elastic cord.
b) Determine Young's modulus (E) for the cord.
c) I found $x = 1.0$ cm for a mass of 100.0 g and an initial length of 0.80 m. Get numerical values for each of your above results.

12.30 As the cord stretches its cross-section decreases. This phenomenon is called the Poisson effect. I made a student (Andrew) repeat the above experiment with different weights on a 1.00 m cord, simultaneously measuring the elongation and the diameter of the cord. He tabulated the data and made the graph shown below.

a) Compare the Young's modulus equation to the graph. Is the slope E or $1/E$?
b) Does the slope change as load increases?

m (g)	x (cm)	D (mm)	x/L	F/A (MPa)
50	0.5	2.7	0.005	0.21
100	1.0	2.5	0.010	0.50
150	2.0	2.4	0.020	0.81
200	3.3	2.3	0.033	1.18
250	6.5	2.3	0.065	1.47
300	11.5	2.2	0.115	1.93
350	16.5	2.1	0.165	2.48
400	22.0	2.1	0.220	2.83
450	27.5	2.0	0.275	3.51
500	32.5	2.0	0.325	3.90

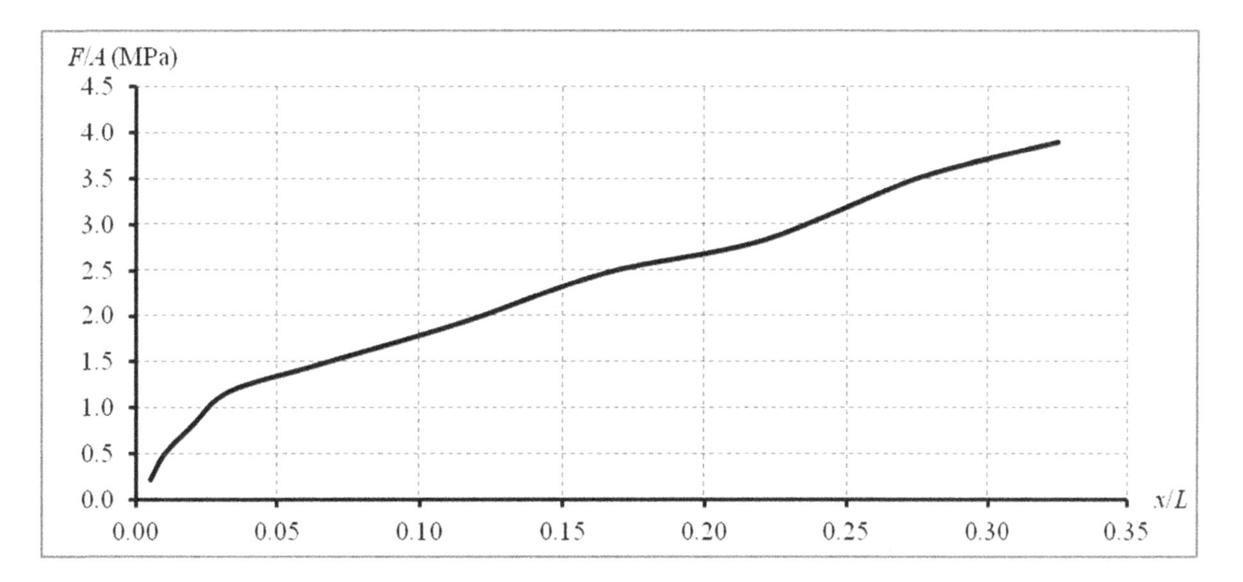

The Poisson effect relates to a topic in engineering called necking…look it up. It is interesting to note that some materials, sometimes called auxetic materials, exhibit the opposite effect. When they are stretched the cross section tends to increase! How bizarre! Humans are now creating materials not found in nature (called metamaterials). These metamaterials have bizarre properties that have novel applications. Some obey left-hand rules instead of right-hand rules! If this type of stuff interests you, materials science is for you.

Last note: engineering stress-strain curves assume constant cross-sectional area is used. This makes the graphs appear to have negative slopes in certain regimes. You can discuss this with your engineering instructor.

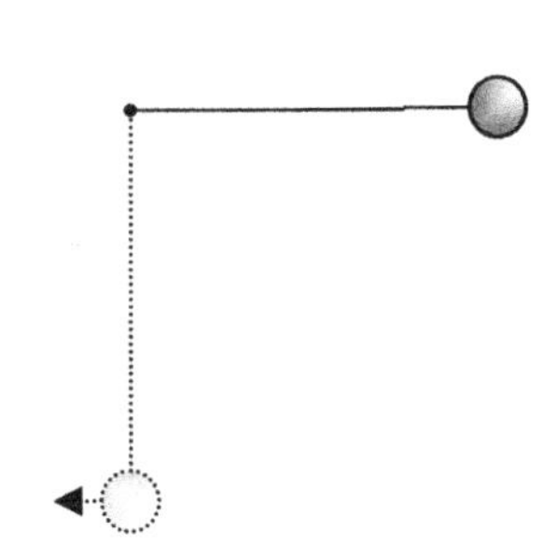

12.31 A bowling ball of mass m and radius R is attached to the ceiling with a long wire. The wire has length L and diameter D. The ball is released from rest when the string is parallel to the floor. Young's modulus of the material is E.

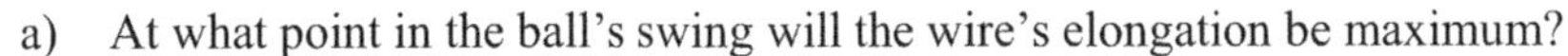

a) At what point in the ball's swing will the wire's elongation be maximum?

b) What is the speed of the ball at the lowest point in the swing? For this part, ignore any stretch in the wire. Answer in terms of L, R, and g.

c) What is the tension in the string in terms of only m and g.

d) Determine the fractional elongation ($\Delta L/L_0$) at the lowest point in the swing? Answer: $\frac{12mg}{\pi D^2 E}$.

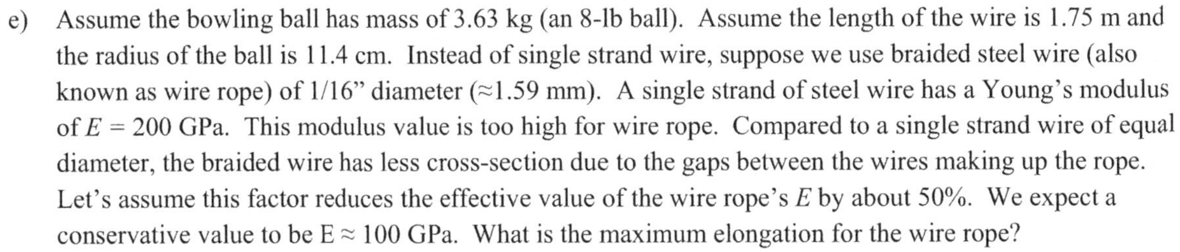

e) Assume the bowling ball has mass of 3.63 kg (an 8-lb ball). Assume the length of the wire is 1.75 m and the radius of the ball is 11.4 cm. Instead of single strand wire, suppose we use braided steel wire (also known as wire rope) of 1/16" diameter (≈1.59 mm). A single strand of steel wire has a Young's modulus of E = 200 GPa. This modulus value is too high for wire rope. Compared to a single strand wire of equal diameter, the braided wire has less cross-section due to the gaps between the wires making up the rope. Let's assume this factor reduces the effective value of the wire rope's E by about 50%. We expect a conservative value to be E ≈ 100 GPa. What is the maximum elongation for the wire rope?

f) **Challenge:** Verify it is reasonable to ignore the stretch in the wire (rope) for part b. Determine the % change in speed if the elongation of the wire (rope) is included. I used the binomial expansion to show $\%\Delta v \approx \frac{\Delta L}{2(L+R)} \times 100\% = 0.013\%$

g) **Going Further:** Suppose the ultimate strength of the wire (rope) is approximately 200 MPa. How does the loading of the wire (rope) at the bottom of the swing compare to the ultimate strength? You might need to do a bit of web research to determine what ultimate strength means and to estimate a value. Ultimate strength is a topic discussed in a materials class.

UNIVERSAL GRAVITATION

As an object rises above the earth, the gravitational force acting changes with altitude. From a common sense standpoint, we expect the object will feel no gravitational force once it is very far from earth. Newton determined an equation for gravitational force between any two objects. He found the gravitational force object 1 exerts on object 2 is given by

$$\vec{F}_{\mathbf{1on2}} = \frac{Gm_1m_2}{r_{1to2}^2}(-\hat{r}_{1to2})$$

where $G = 6.67 \times 10^{-11}\,\frac{\text{N}\cdot\text{m}^2}{\text{kg}^2}$ and $\vec{r}_{1to2}$ is a displacement vector pointing from the center of 1 to the center of 2. Here are a few key points worth summarizing:

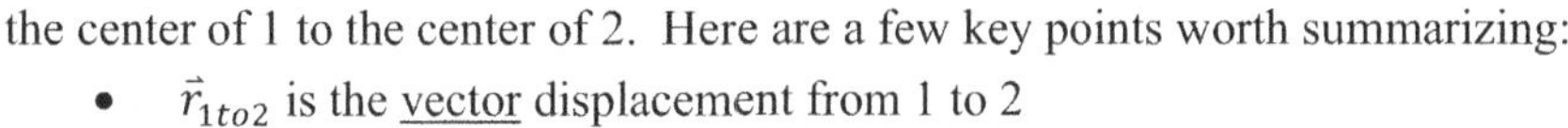

- $\vec{r}_{1to2}$ is the vector displacement from 1 to 2
- r_{1to2} is the scalar distance (always positive)
- Remember r_{1to2} = center-to-center distance
- $\hat{r}_{1to2} = \frac{\vec{r}_{1to2}}{r_{1to2}}$ and points from 1 to 2 (from what is *causing* the force to what is *experiencing* the force)
- The magnitude of the gravitational force is $F_{1on2} = \frac{Gm_1m_2}{r_{1to2}^2}$
- Mass 1 attracts mass 2… $\vec{F}_{\mathbf{1on2}}$ is directed towards mass 1 ($\vec{F}_{\mathbf{1on2}}$ points $-\hat{r}_{1to2}$)

From this force equation one can also determine the gravitational potential energy associated with a pair of masses. The gravitational potential energy associated with this conservative force is given by

$$U_{12} = -\frac{Gm_1m_2}{r_{12}}$$

Here are a few key points worth summarizing:

- The potential energy equation is associated with a pair of masses (not a single mass). The two masses share the potential energy associated with their gravitational fields.
- When using potential energy, no vectors are required.
- Potential energy is always negative
- Potential energy uses $\frac{1}{r_{12}}$ while force uses $\frac{1}{r_{12}^2}$

13.1 During a solar eclipse the earth, sun, and moon are aligned as shown in the figure (not to scale). Assume the earth has mass 5.97×10^{24} kg and radius 6.37×10^{6} m. The sun has mass 1.99×10^{30} kg and radius 6.96×10^{8} m. The moon has mass 7.35×10^{22} kg and radius 1.74×10^{6} m. The center-to-center distance from earth to sun is 1.50×10^{11} m while the center-to-center distance from earth to moon is 3.84×10^{8} m.

a) Determine the magnitude of the gravitational force exerted by the sun on the earth.
b) Determine the magnitude of the gravitational force exerted by the moon on the earth.
c) Determine the potential energy associated with the earth and sun.
d) Determine the potential energy associated with the earth and moon.
e) If we add the previous two results, we do not have the total gravitational potential energy stored in the system. What is the total gravitational energy of the system?
f) The sun exerts a much larger force on the earth yet the moon has a greater influence on tides. Why?

13.2 Assume the earth has $M = 5.97 \times 10^{24}$ kg and radius $R = 6370$ km. Consider a mass with m above the earth's surface. Rather than discussing the center-to-center distance, it is common to describe the altitude of m. Altitude (h) is the distance above the earth's surface. Notice the magnitude of the force can be expressed as

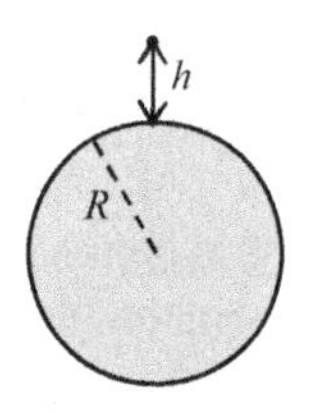

$$F_G = \frac{GMm}{(R+h)^2}$$

This result is identical to $F = mg$ if we let the magnitude of the acceleration due to gravity at altitude h be

$$g = \frac{GM}{(R+h)^2}$$

Create a table of h, F, and U for $h = 0$ to $3R \approx 20000$ km. Use $m = 1$ kg. This is convenient because the numerical value of the force is equal to the *effective* numerical value of g for any given altitude. Use the table to make plots of F vs h and U vs h. This is a worthwhile exercise to ensure you are using prefixes and scientific notation properly. Better to find your mistakes now than on test day!

13.3 Consider a satellite of mass m in a circular orbit of altitude h around a planet of mass $M$and radius R. If we assume $M \gg m$ we can consider the planet as stationary and use the center of the planet as the origin of our coordinate system. Figure not to scale.

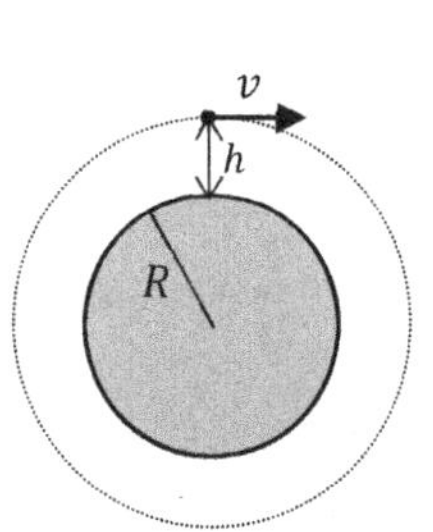

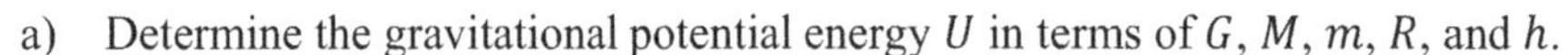

a) Determine the gravitational potential energy U in terms of G, M, m, R, and h.
b) Determine the speed of the satellite in terms of G, M, m, R, and h. Hint: do an FBD.
c) Determine the kinetic energy K of the satellite. Compare K to U.
d) Write down the total energy E of the orbit. Is it positive or negative?
e) If you go to a higher orbit does energy increase or decrease?
f) Determine the period $\mathbb{T}$ of the orbit. This is Kepler's 3rd Law.
g) A CubeSat is a micro-satellite with a mass of about 1 kg. Assume the earth has $M = 5.97 \times 10^{24}$ kg and radius $R = 6370$ km. Create a table of h, v, $\mathbb{T}$, and E for $h = 0$ to $6R$. Plot of v, $\mathbb{T}$, and E versus h. I used km for h, km/s for v, hours for $\mathbb{T}$, and MJ for E.

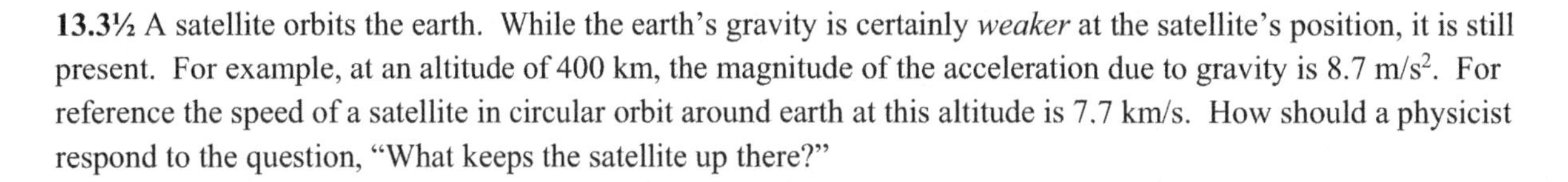

13.3½ A satellite orbits the earth. While the earth's gravity is certainly *weaker* at the satellite's position, it is still present. For example, at an altitude of 400 km, the magnitude of the acceleration due to gravity is 8.7 m/s^2. For reference the speed of a satellite in circular orbit around earth at this altitude is 7.7 km/s. How should a physicist respond to the question, "What keeps the satellite up there?"

13.4 Consider doing a simulation of point masses. A suggested simulation is available on www.robjorstad.com in the supplemental worksheets. Even better, code up your own simulation…it is not as bad as you think.

13.5 The figure shows a top view of three masses at rest on a table. The masses m_1, m_2, & m_3 are 1.00, 2.00, & 3.00 kg respectively. The center-to-center distance between m_1 and m_2 is 30.0 cm; center-to-center distance between 1 and 3 is 40.0 cm. Assume the origin lies at the center of m_3. In solutions I used green for 1 on 3 and red for 2 on 3.

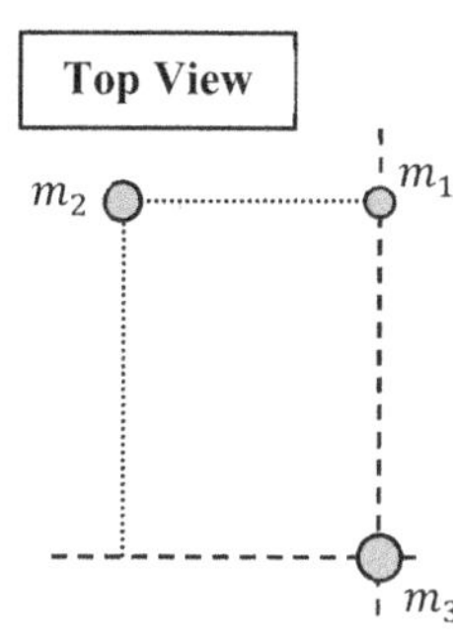

a) Before doing math, take a guess: which mass, m_2 or m_1, exerts a larger force <u>on</u> m_3? In which direction does the net gravitational force exerted <u>on</u> m_3 <u>by</u> m_2 & m_1 point?
b) Determine the net gravitational force exert <u>on</u> m_3 <u>by</u> m_2 & m_1. Express your result as a magnitude and direction.
c) Assuming these masses are on a table near the earth's surface. How much larger is the earth's gravitational force compared to the gravitational force exerted <u>on</u> m_3 <u>by</u> m_2 & m_1?
d) Determine the gravitational potential energy associated with this group of masses.

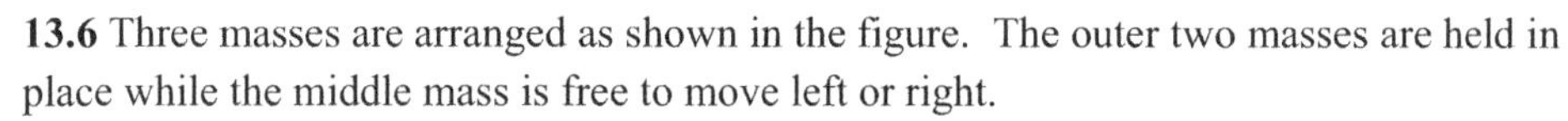

13.6 Three masses are arranged as shown in the figure. The outer two masses are held in place while the middle mass is free to move left or right.

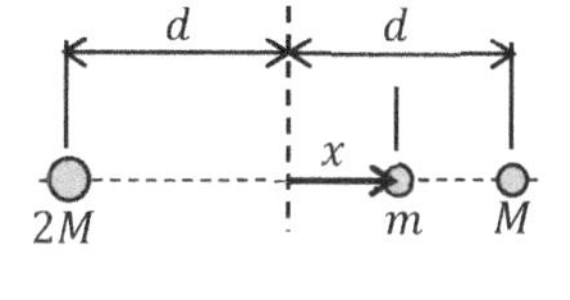

a) Determine the potential energy U in terms of G, M, m, r, and x for $-2d < x < 2d$.
b) Determine the net gravitational force acting on the middle mass for $-2d < x < 2d$.
c) At what location is the middle mass in equilibrium?
d) Assume $M = 10^{16}$ kg, $m = 10^{15}$ kg, and $d = 10^{11}$ m. Plot F_x vs x for $-2d < x < 2d$.
e) Using the same parameters, plot U vs x for $-2d < x < 2d$.

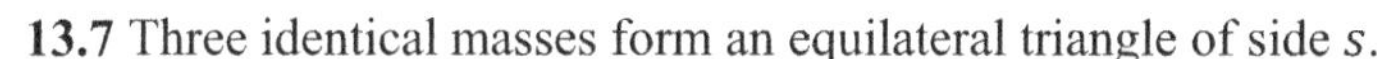

13.7 Three identical masses form an equilateral triangle of side s.

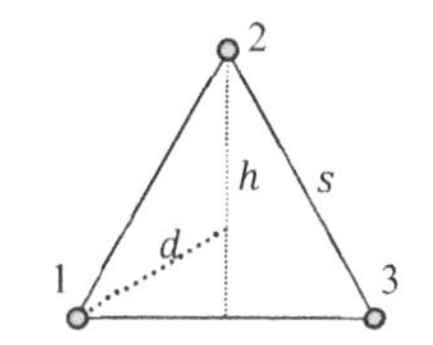

a) Determine the height h of the triangle in terms of s.
b) Determine the distance d from any corner to the center of the triangle. Notice $d \neq \frac{h}{2}$.
c) Determine the gravitational potential energy of the masses.
d) Suppose a fourth identical mass is brought to the center. What is the <u>change</u> in gravitational potential energy of the system?
e) Determine work done by gravity as the fourth mass is brought from far away to the center of the triangle.
f) Determine the net gravitational force on the fourth mass at the center.
g) Suppose the fourth mass was instead brought to a point midway between masses 1 and 3. Determine the net force on the fourth mass at *this* location.

Notice that symmetry, if applicable, can make force problems *much* easier.

13.8 Four identical masses held in place on the corners of a square of side s.

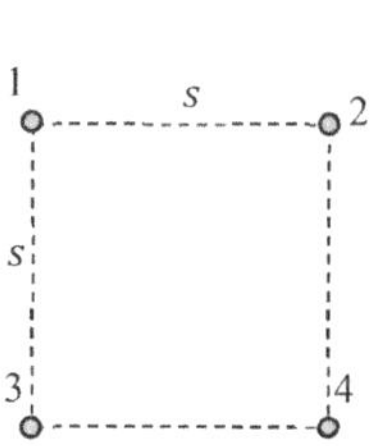

a) Determine the net gravitational force on mass 4.
b) Determine the gravitational potential energy stored in the system.
c) Now suppose mass 4 is released from rest and travels towards mass 1. The others are still held in place. Determine the net force on 4 when it reaches the center of the square.
d) Determine the GPE when mass 4 is at the center of the square. Think: should GPE increase or decrease? Should it become more negative or more positive?
e) Is the work done by gravity positive or negative as mass 4 is moved to the center of the square?
f) Explain why mass 4 travels directly towards 1 and not on some other path.
g) Determine the speed of mass 4 as it reaches the center. Answer as a number with 3 sig figs times $\sqrt{\frac{Gm}{s}}$.
Think: why can you <u>not</u> use the result of part a to get an acceleration and do kinematics?
h) If <u>one</u> of the masses is doubled, will mass 4 still travel towards mass 1? What if <u>two</u> of the masses are doubled?

13.9 Escape from planet of the physicists

Far into the future, a planet seems eerily similar to earth except physicists are in charge of everything. Everything works ok but only to within 10% of predictions. People assume everything is a sphere all the time. Social interaction is always more awkward than it should be while personal hygiene is never what it could be. Instead of watching TV people constantly plot nefarious equations into graphs to create, in theory, better toasters.

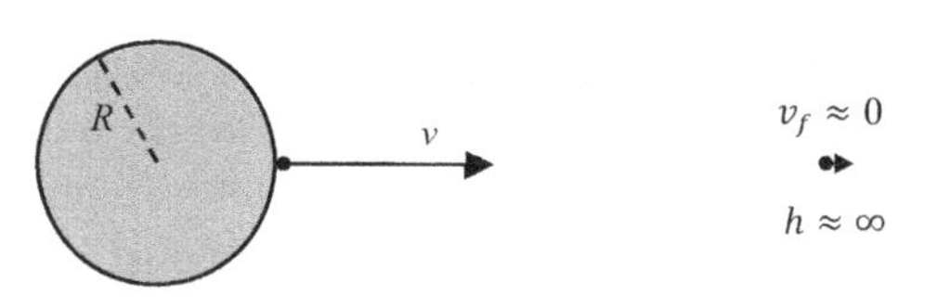

The mass of the planet is M and the radius is R. Fed up with all the nonsense, a few sane individuals seize control of a rail gun used to launch space pods. The people squeeze into the space pod and are launched from the surface at speed v. They don't really care where they are going as long as it is away from the planet of the physicists. The rail gun has negligible length compared to R. After leaving the rail gun, assume drag is negligible and the only force acting on the space pod is gravity. To escape the planet they must be able get enormous ($h \approx \infty$) altitude while still having at least a tiny bit of speed ($v_f \approx 0$).

a) Determine the initial launch speed required to escape the planet.
b) Wait a minute...this planet is earth with $M = 5.97 \times 10^{24}$ kg and radius $R = 6370$ km. Determine a numerical value for the escape speed.
c) The muzzle velocity of an AK-47 is about $700\ \mathrm{m/s}$. If someone fires bullets from an AK-47 straight up will they escape or come back down to earth?
d) Why is it often advantageous to launch satellites from launch sites as close to the equator as possible? Note, the major space launch sites in the US are in southern locations like Texas, Florida, and LOMPOC.
e) Black holes are objects so dense not even light, with speed $c = 3 \times 10^8\ \frac{\mathrm{m}}{\mathrm{s}}$, can escape. The formula derived happens to be correct for determining the event horizon of non-rotating black holes. If light is emitted from an atom at some radius less than the event horizon, the light will not escape...

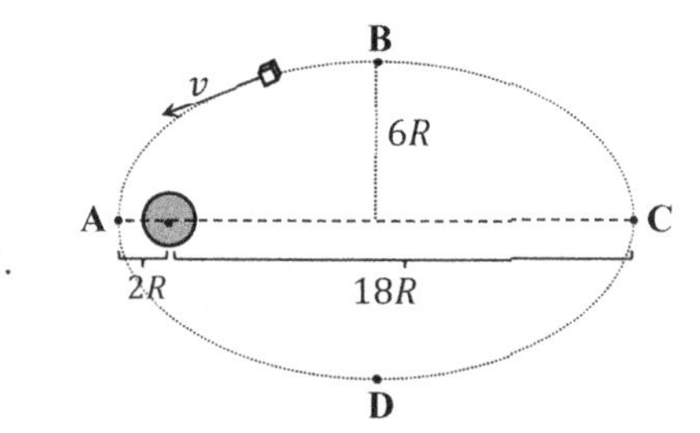

13.10 A 1.00-kg CubeSat orbits a planet of mass $M = 5.97 \times 10^{24}$ kg and radius $R = 6370$ km. The figure at right is to scale (except the size of the cube sat).
The perigee (closest approach) is distance $2R$ at point **A**.
The apogee (farthest distance) is distance $18R$ occurring at point **C**.
The semi-major axis of the ellipse is $a = 10R$ while the semi-minor axis is $b = 6R$.
The planet is at a focal point of the ellipse.

a) Can we use conservation of energy to relate points in the orbit? Explain.
b) Can we use conservation of angular momentum? Explain.
c) Fill in the table below with numerical values. Hint: do *not* do circular motion force problems for non-circular orbits. Use conservation of energy, angular momentum, or both. It is tricky to get started so I figured out one velocity for you.

Point	r (units of R)	F (N)	$r_\perp$ (units of R)	L (kg·km²/s)	v (km/s)	K (MJ)	U (MJ)	E (MJ)
A	2	2.45			7.51			
B								
C								

13.11 Consider a mass m initially located distance r_i from the center of mass M. The mass m moves radially outward to a new radius r_f. Recall that force, work, and potential energy are related by the equation

$$\Delta U = -W = -\int_i^f \vec{F} \cdot d\vec{s}$$

a) For a particle moving radially outwards $d\vec{s} = dr\hat{r}$. Determine ΔU of the system as m moves from r_i to r_f.

b) Now set $r_i = R$ and $r_f = \infty$. Determine the potential energy of the pair of masses when $r_i = R$. Hint: it is reasonable to assume that $U = 0$ when $r_f = \infty$. Compare your result to $U_{12} = -\frac{Gm_1m_2}{r_{12}}$.

13.12 In Chapter 8 we assumed assuming $U = mgh$ where h is the altitude above the earth's surface. This seems very different from $U = -\frac{GmM_E}{R_E+h}$. For these two equations to produce consistent predictions they must give identical results for *changes* in potential energy (ΔU) when h_i & $h_f \ll R_E$.

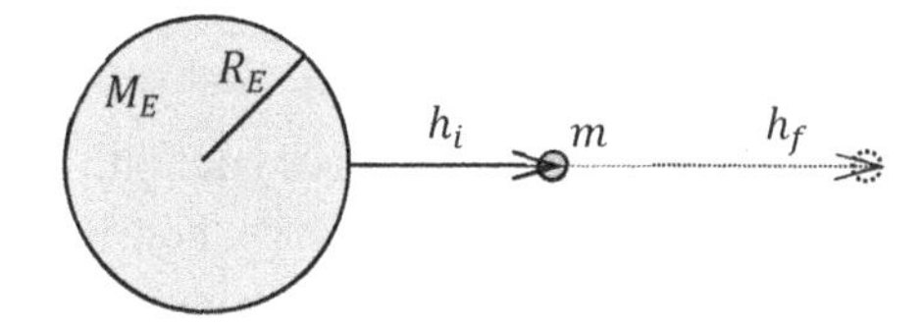

a) Show U can be rewritten as $U = -\frac{GmM_E}{R_E}(1+\delta)^{-1}$ where $\delta = \frac{h}{R_E}$.

b) If $h \ll R_E$ we expect $\delta = \frac{h}{R_E} \ll 1$. We may then use the binomial expansion or Taylor series expansion to give a good approximation for U. The binomial expansion for $(1+\delta)^n \approx 1 + n\delta$. Use the binomial expansion to show $U \approx -\frac{GmM_E}{R_E}\left(1 - \frac{h}{R_E}\right)$.

c) Show $\Delta U = mg(h_f - h_i)$ when h_i & $h_f \ll R_E$. This shows that $U = mgh$ and $U = -\frac{GmM_E}{R_E+h}$ produce consistent predictions near earth's surface.

Why care? I think the binomial expansion trick is really important/cool. It is used by scientists and engineers to turn really ugly equations into linear ones. As another example, the speed of sound in air is given by

$$v = \sqrt{\frac{\gamma RT}{M}} = \sqrt{\frac{\gamma R(273 + T_C)}{M}}$$

where γ is the adiabatic constant, R is the ideal gas constant, T is the temperature of the gas in Kelvin, and T_C is the temperature in Celsius. Using a binomial expansion the formula looks like

$$v = 331.4 + 0.6T_C$$

The binomial expansion makes an annoying equation slightly less painful to use. Expansions are used in thermodynamics, electricity and magnetism, special relativity, quantum mechanics, and more. Use of expansions (binomial or Taylor series) is a bit like the quadratic formula; you don't use it every day but it is an important tool to get the job done efficiently on certain occasions. A few more problems in this chapter use them so you can practice.

13.13 Three masses are arranged as shown in the figure. Note: this one uses all little m's and x is defined differently than in **13.6**.

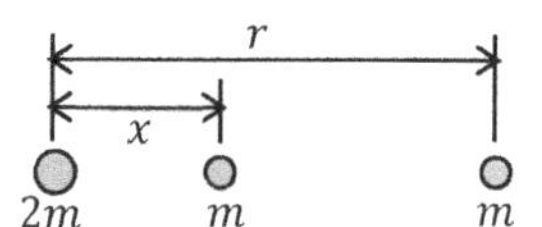

f) Determine the potential energy U in terms of G, m, r, and x.

g) Determine the net gravitational force acting on the middle mass.

h) Previously we learned that $F_x = -\frac{dU}{dx}$. Assuming r is constant, take the derivative of your result for part a and compare it to part b. Is $F_x = -\frac{dU}{dx}$ valid for this situation?

i) How could you use U to determine the force in the x-direction for the rightmost mass?

j) **Challenge:** how could you determine the force on the leftmost mass?

Notice it is fairly easy to write down the energy equation. Software exists for taking symbolic derivatives. One method to determine the force on a particular mass in a complicated problem is to write down U algebraically and use software to take the derivatives. When using this method for a complicated 3D problem people are less likely to make mistakes compared to getting the signs correct on multiple 3D vectors!

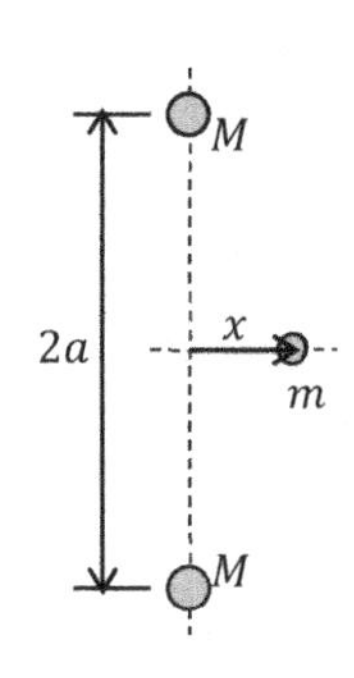

13.14 Two objects of mass M are distance a above and below the origin as shown in the figure. A third mass m lies distance x to the right of the origin.

a) Determine the potential energy of the three mass system.
b) Determine the net gravitational force on the middle mass.
c) What value of x gives a maximum force? What is the maximum value of the force?
d) Assuming $x \ll a$, use the binomial expansion to determine an approximate equation for the force.
e) Plot F_x vs x assuming $M = 10^{24}$ kg, $m = 1$ kg, and $a = 10^8$ m. Use values of x ranging from $-3a$ up to $+3a$. This is approximately like a 1kg satellite halfway between two earth's separated by the same distance as the earth and moon.

Notice if we keep only the first term in the expansion $\vec{F} = k_{eff}x(-\hat{\imath})$ where $k_{eff} = \frac{2GMm}{a^3}$. This implies the center mass should oscillate back and forth as if it were attached to a spring. Notice the plot of F_x vs x is linear, just like a spring, when $x \ll a$. Notice, however, the situation is unstable. If mass m is slightly above or below the horizontal axis, the vertical forces no longer cancel out; the middle mass is pulled towards the closer of the big masses.

13.15 Two 100 kg cubes of gold rest on horizontal, level surface (near earth). The density of gold is 19.3 g/cm^3. The side faces of the blocks parallel to each other and are 10 cm apart.

a) Determine length of one side of cube using density.
b) Determine minimum coefficient of friction to keep blocks from sliding.
c) Is it realistic to expect the blocks of gold to slide towards one another even though, strictly speaking, they are attracted to each other by a gravitational force? Can you use gravitational force to attract other people to you?

13.16 Consider a satellite or mass m in a circular orbit of radius R around a central mass M. From astronomical observations, the period of the earth's orbit around the sun is known to be 365.25 days and the center-to-center distance from earth to sun is 1.50×10^{11} m. Determine the mass of the sun using an FBD and force equation.

13.17 What altitude is required for a satellite to be in geo-synchronous orbit around earth? A geo-synchronous orbit implies the period of the orbit matches the period of the earth's spin rate. Assume the earth has mass 5.97×10^{24} kg and radius 6.37×10^6 m.

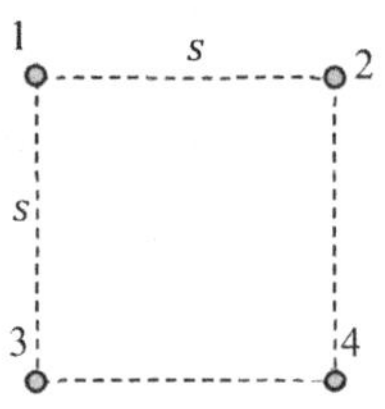

13.18 Four identical spheres, each mass m, are held in place at the edge of a square of side s. This time all four are released from rest.

a) Determine the speed of each mass after each has traveled 1/3 of the distance to the center. Answer in terms of G, m, and s.
b) Determine the ratio of the initial force on each mass to the force on each mass after each has traveled 1/3 of the distance to the center.
c) Describe how the acceleration changes as the particles move towards the center. In particular, does *acceleration* (magnitude) increase, decrease, as it moves towards the center? Does the *rate of change in acceleration* (a.k.a. jerk) increase, decrease, or remain constant?

13.19 In problem **13.6** masses were arranged as shown in the figure. The outer two masses were held in place while the middle mass is free to move left or right. The gravitational potential energy of the system was given by

$$U = -GM\left(\frac{2m}{d+x} + \frac{m}{d-x} + \frac{M}{d}\right)$$

Note: I am assuming $-d < x < d$ so I can drop the absolute value signs for this part.
The horizontal force on the middle mass was found to be

$$F_x = GMm\left[\frac{1}{(d-x)^2} + \frac{-2}{(d+x)^2}\right]$$

a) What ratio of parameters must be small to effectively use a binomial expansion *near* the origin?
b) Use a binomial expansion to determine approximations for F_x and U *near* the origin.
c) How do the previous results change if you desire approximate formulas for locations *far* from the origin?

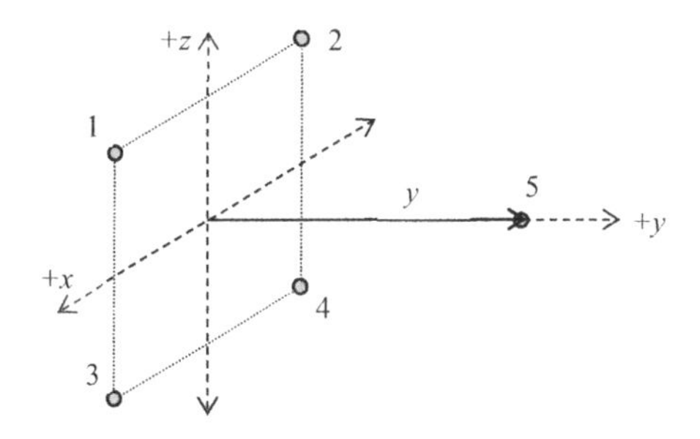

13.20 Four identical masses are arranged on a square of side s that lies in the xz-plane as shown in the figure. A fifth identical mass is located distance y from the origin on the y-axis.

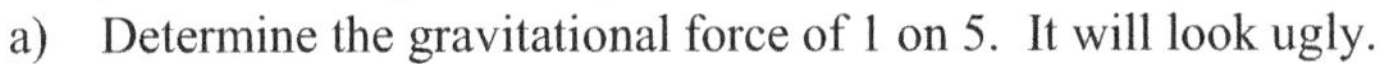

a) Determine the gravitational force of 1 on 5. It will look ugly.
b) Think: what direction is the *net* gravitational force on 5? Hint: symmetry.
c) Determine the net force on 5. Isn't symmetry awesome!

13.21 Three point masses are arranged as shown in the figure. The distances are noted as multiples of the arbitrary distance d. You may assume d has units of meters.

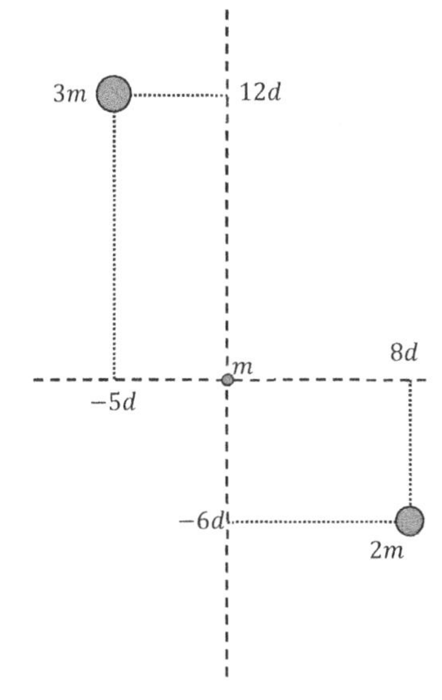

a) Determine the gravitational potential energy associated with this group of point masses. Answer with a 3 digit decimal times $\frac{Gm^2}{d}$.
b) Determine the *magnitude* of the net gravitational force exerted on the mass at the origin. Answer with a 3 digit decimal times $\frac{Gm^2}{d^2}$.
c) Determine the *direction* of the net gravitational force exerted on the mass at the origin. Express your final answer in degrees with three sig figs. Include a sketch showing the angle to help clarify your answer.

Reduced mass

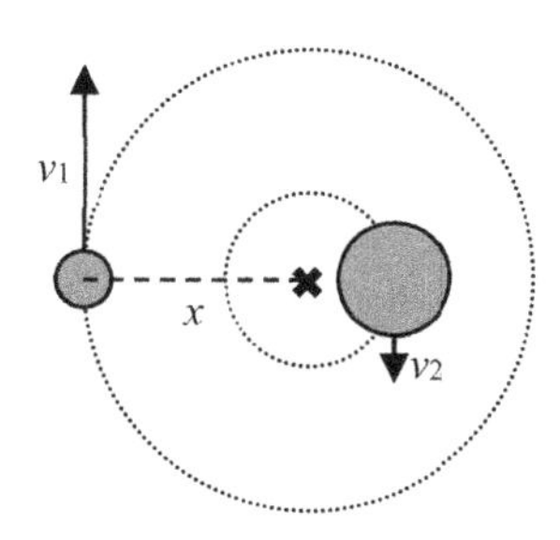

In all of the previous orbit problems, we have been assuming the mass m of the satellite is much smaller than the central mass M. If this assumption is not valid, the central mass cannot be considered as at rest. The radius of circular motion is not the center-to-center distance R. Both M and m will be in circular orbits about the center of mass! Since the two objects are orbiting with different radii (and speeds) it will be less confusing to use $a_c = r\omega^2$instead of $a_c = \frac{v^2}{r}$. The two must have the same ω. Think: the two masses are always opposite each other on the circle, they must have the same rotation rate.

a) Show the distance x from the center of the m to the center of mass of the system is $x = R\frac{M}{(m+M)}$.

b) Write down the force equation for the circular motion of m using R as the center-to-center distance but x as the radius of circular motion. You should find

$$\frac{mM}{(m+M)}R\omega^2 = \mu R\omega^2 = \frac{GmM}{R^2}$$

where the *reduced mass* is defined as $\mu = \frac{mM}{m+M}$. Notice the force still uses the entire mass m while the acceleration is reduced using the reduced mass trick.

c) Assume the earth has mass 5.97×10^{24} kg and radius 6.37×10^6 m. The moon has mass 7.35×10^{22} kg and radius 1.74×10^6 m. The center-to-center distance from earth to moon is 3.84×10^8 m. Determine x and μ. I found μ =7.35×10^{22} kg and x = 3.80×10^8 m. It is fun to notice the center of mass of the system is about 4×10^6 m from the center of the earth. This is approximately 70% of the earth's radius.

Note: a similar situation occurs an electron orbits a hydrogen nucleus. The spectra of light produced relates to the energies of the orbits. The reduced mass must be used to correctly predict the energies of orbits and the colors of light produced in the spectra! I find this interesting.

The tides

Trying to understand the tides is no joke. A lot of factors play into it. The dominant factor influencing the tides is the moon so let's start there.

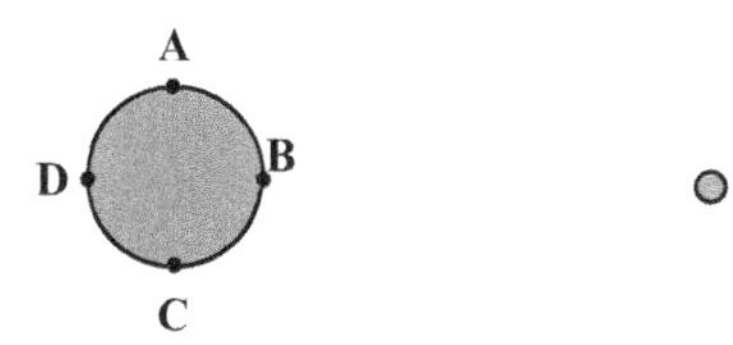

Consider one instant in time where the moon is exerting forces on the earth. A figure, not to scale shows the two objects. Four points on the earth are indicated by the black dots.

a) For each points, sketch the direction of the gravitational force the moon would exert on a 1 kg chunk of water at the earth's surface. For reference I found the magnitudes at **A** & **C** as 97% of the magnitude for **B**. Point **D** has about 93% of the magnitude of **B**.

b) Assume the earth-moon distance is D and the radius of the earth is R. Assuming $R \ll D$. Comparing points **A** and **B**, show difference in force exerted by the moon on mass m

$$\Delta F = \left|\left(-\frac{GmM}{(D-R)^2}\right) - \left(-\frac{GmM}{(D+R)^2}\right)\right| \approx 4\frac{GmM}{D^3}R$$

Hint: use a binomial expansion. The difference in force is proportional to distance cubed not squared! This helps explain why the sun with a much larger force plays a lesser role in influencing the tides.

Factors influencing the tides include:

- The moon's gravitational force the ocean varies slightly in magnitude and direction depending on location. This gives rise to the differential force discussed above.
- From the reduced mass problem we know the earth is rotating about the earth-moon center of mass and not simply spinning in place.
- The oceans form tidal bulges as they are pulled by the moon's gravitational forces but the continents do not. The tidal bulges of water trying to rotate westward relative to the earth's surface are impeded by the eastern coastlines of the continents
- The moon's orbit is elliptical. The earth moon distance ranges from about 3.6×10^8 m to 4.1×10^8 m.
- The plane of the moon's orbit is angled relative to a plane through the earth's equator.
- The sun affects the tides about half as much as the moon.
- Coastline patterns, and wind and weather all contribute to the tides as well.
- Islands in the middle of the ocean typically experience small tidal differences (less than a meter) but the Bay of Fundy in Nova Scotia can occasionally experience tides huge tidal difference ($>$ 15 m $\approx$ 50 ft). Some places, like the Gulf of Mexico, only experience one tide a day instead of two. In others there are effectively no tides per day.

A fun resource with easy reading can be found by a web search for "NOAA tides education".

Is it possible for an astronaut to accidentally jump off the moon and launch herself into space?

What is escape velocity on the moon?

Determine the take-off speed of an extreme athlete; we expect that on earth an extreme athlete might have a max vertical jump height of about 1.3 m. Suppose an astronaut with all gear could have a max vertical jump (om earth) of about half that.

Upon comparing the two speeds, do you think an astronaut needs to worry about accidentally jumping off of the moon?

Assume an asteroid has the same density as earth. Assume it is a sphere of radius r. What radius is required for the astronaut to be able to jump off the asteroid and launch into space?

PRESSURE & FLUIDS

$$\text{Pressure} = \frac{\text{Force}}{\text{Area}} \qquad P = \frac{F}{A} \qquad \text{Units of Pressure: } 1 \text{ Pascal} = 1 \text{ Pa} = 1 \frac{\text{N}}{\text{m}^2}$$

For physics classes it is often convenient to express standard atmospheric pressure as

$$P_0 = 1.013 \times 10^5 \text{Pa} \approx 1.0 \times 10^5 \frac{\text{N}}{\text{m}^2} \qquad P_0 = 14.696 \frac{\text{lbs}}{\text{in}^2} \approx 15 \text{ PSI}$$

Notice the units of force appear in each of these units.

Standard atmospheric pressure is also expressed as

$$P_0 = 1 \text{ atm} \qquad P_0 = 760 \text{ torr} \approx 760 \text{ mm Hg}$$

The units of mm Hg relate literally to the height of a mercury column in a Torricelli barometer shown at right. The barometer is made by inverting a tube of mercury and placing it into a dish. Gravity pulls the mercury downwards leaving negligible air pressure above the mercury column. Since atmospheric pressure is pushing down on the dish, the height of the column rises and falls with atmospheric pressure. If the height of the mercury column is 760 mm tall, the atmospheric pressure is said to be standard.

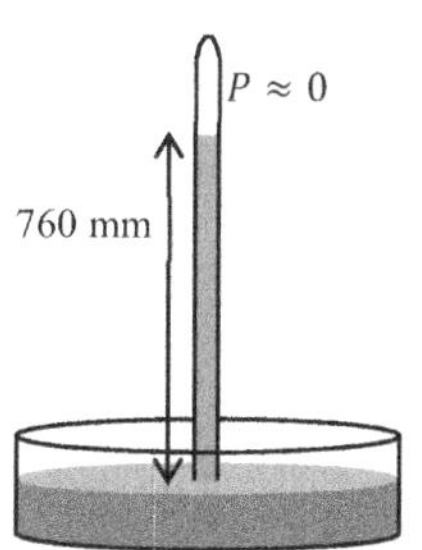

Later mm of Hg were renamed torr. When 1 torr was redefined as 1/760th of the standard atmospheric pressure it no longer exactly equated to 1 mm of Hg. For all intents and purposes, the two units are interchangeable (unless many sig figs are required).

14.1 Susie Smith is doing push-ups. She does the first five with flat palms and the last five on her fingertips.

a) Which push-up, if any, causes the floor to exert a greater <u>force</u> on the human?
b) Which push-up, if any, causes greater <u>pressure</u> on the human (for the body part in contact with the floor)?

14.2 Billy Bob stands on a scale with both feet. The scale reads 200 lbs. Billy Bob raises his right foot and stands motionless with his left foot still on the scale. Which of the following change? Which remain constant?

Force of scale on Billy's left foot	Net force of scale on Billy	Pressure exerted by scale on Billy's left foot

14.3 Snowshoes are essentially big flat paddles strapped to the bottom of your boots or shoes. The snowshoes have an area much larger than the boots to which they are affixed. From a force/pressure standpoint, explain why putting on snowshoes is advantageous for trying to walk on snow.

14.4 Pressure differences cause a net force from high to low pressure.
A building has a flat roof that is 10.0 m by 15.0 m. Assume all doors & windows on the building are sealed tight when the air *inside* is at normal atmospheric pressure. Suddenly, a violent storm causes atmospheric pressure *outside* the building to drop by 15%. This dramatic pressure drop *might* be possible during a tornado. The roof has a mass of 18×10³ kg (approx. 20 tons).

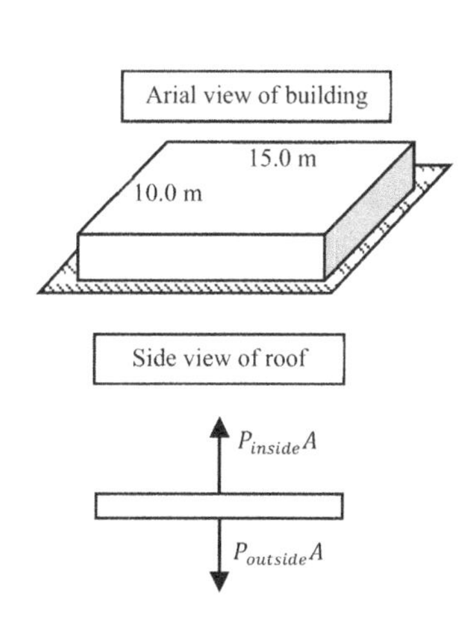

a) Determine <u>upwards</u> force on the roof caused by air <u>inside</u> the building.
b) Determine <u>downwards</u> force on the roof caused by air <u>outside</u> the building.
c) Determine the <u>net</u> force caused by this <u>pressure difference</u>.
d) Compare the weight of the roof to the force from the pressure difference. Do the walls of the building need to exert an upwards or a downwards force to keep the roof in equilibrium?

Note: in real life, air flows complicate this vastly over-simplified scenario.

How high could *water* rise in a vacuum?

14.5 Mercury rises 760 mm in a pure vacuum. The density of mercury is 13.5 times greater than water.

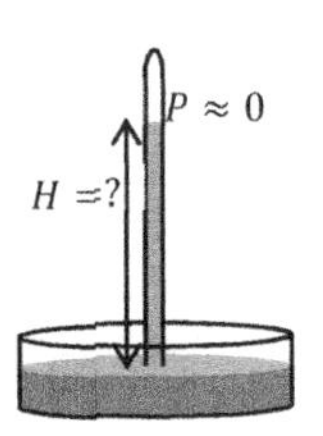

a) How tall would the glass tube need to be if Torricelli had used water instead of mercury?

b) Think about the pros and cons of making such a barometer out of water.
Would scale readings be more or less precise?
Would the water evaporate faster or slower than the mercury?
Where would you put the thing & how would you reach the top of the column to read the scale?

c) Aren't you curious, then, how water can travel upwards from the root system of a plant to the top of a giant redwood tree which can grow over 100 meters tall? Do an internet search for "trees taller than 10 m".

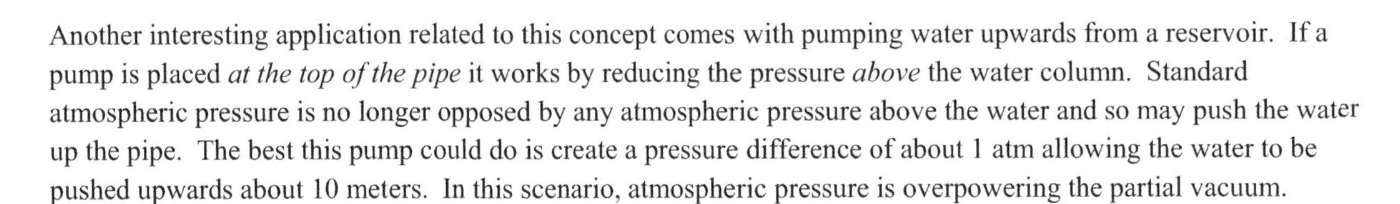

Another interesting application related to this concept comes with pumping water upwards from a reservoir. If a pump is placed *at the top of the pipe* it works by reducing the pressure *above* the water column. Standard atmospheric pressure is no longer opposed by any atmospheric pressure above the water and so may push the water up the pipe. The best this pump could do is create a pressure difference of about 1 atm allowing the water to be pushed upwards about 10 meters. In this scenario, atmospheric pressure is overpowering the partial vacuum.

Different types of pumps may be used if the pump is to be placed *at the bottom of the pipe* (by the reservoir surface). In these types of pumps, the water *below* the water column is pressurized beyond standard atmospheric pressure in order to push upwards against standard atmospheric pressure. These types of pumps can easily be designed to exceed standard atmospheres of pressure causing a pressure difference of greater than 1 atm. In this scenario the pump is overpowering atmospheric pressure.

Going further: pump systems can be designed with several pumps in series and parallel for various reasons. Read online if this is your thing…

Incompressibility of water

Strictly speaking, water is compressible. From a practical standpoint, water is often considered incompressible. When a submarine dives beneath the surface of the ocean, the pressure changes dramatically with depth while the change in the density of the water is insignificant. As an example, during a 100 m dive from the surface of the ocean the pressure changes by 10 atm (1000%) but density changes by 0.5 kg/m^3 (0.05%). Since the density of water remains nearly constant despite these large pressure changes, we say water is incompressible.

A dramatic representation the incompressibility of water can be prepared by filling a 1 L spherical glass beaker with water having almost all air bubbles removed. This is done by filling the flask and boiling it for a while to get most of the air out. The flask is then carefully topped off with boiled water as needed and sealed with both a stopper and some silicone sealant. This glass flask can then be used as a hammer if you have excellent aim! When the flask impacts a nail, the water inside is almost completely incompressible. This prevents the glass from flexing and shattering.

It helps to have good aim so you squarely hit the head of the nail; if not it might slip off and smash into the table. It also helps if the nail hasn't gone completely through the board to a harder surface. When you hit a nail into the wood, the time of contact is slightly longer as the nail is driven into the wood. This analogous to throwing an egg at a bedsheet versus at a wall. The bed sheet gives a little extending the collision time and thus reducing the average force during the collision. If you've forgotten all this, review impulse, force, and collision time.

Note: I have let groups of junior high kids go crazy on this thing for several hours and had the flask survive. On the other hand, I've had the flask work for about three people before it shatters. Eventually it always breaks but it sure is fun never knowing who will get soaked by it! Wear appropriate safety protection if you try this demo.

Density Quick Questions

14.6 Don't over think these.

a) Which has greater density, 100 kg of lead or 100 kg of aluminum?

b) Which has greater volume, 100 kg of lead or 100 kg of aluminum?

Pressure vs depth for incompressible, static liquid

A simple equation that describes pressure versus depth for an incompressible, static liquid is given by

$$P(h) = P_{surface} + \rho g h$$

where $P_{surface}$ is the pressure at the surface (often, but not always, atmospheric pressure P_0), ρ is the density of the fluid, g is the magnitude of the acceleration due to gravity, and h is the ***depth*** below the surface. There are several points worth re-emphasizing about this equation:

- This equation is valid for fluids at rest.
- In this equation h is depth (not height). In the Bernoulli equation, later this chapter, the opposite is true!
- This equation is valid for incompressible fluids (density is essentially constant with increasing depth).

14.7 A glass of water is filled to make a water column that is 8.00 cm tall. After a few minutes, a layer of oil 12.0 cm thick poured on top of the water and allowed to come to equilibrium. Assume standard atmospheric conditions apply. I labeled the pressures at each interface to make it easier to communicate with each other. Assume the density of oil is 20% less dense than water.

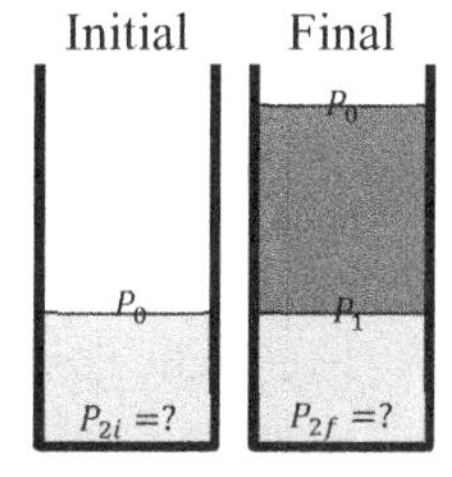

a) Determine the initial pressure at the bottom of the water column.

b) Determine the final pressure at the bottom of the water column.

c) Determine the ratio of the final pressure (with oil) to the initial pressure (without oil).

14.8 A rectangular slab of concrete is used to dam a river and create a reservoir. Two figures showing the important dimensions related to the dam are shown at right. Standard atmospheric conditions apply.

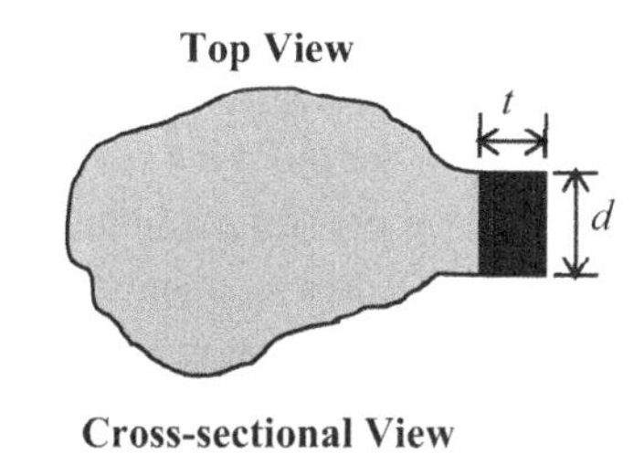

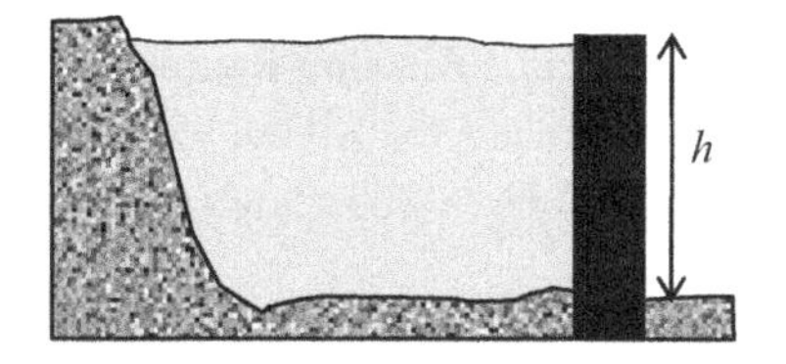

a) Determine the pressure in the water at the bottom of the dam (where the dam meets the bedrock).

b) Since pressure varies linearly with depth for this problem, determine the average water pressure on the dam by averaging the pressure at the top of the dam and at the bottom of the dam.

c) Determine the net force exerted by the water on the dam.

d) Determine the horizontal force exerted by the bedrock to keep the dam in place. Hint: for this problem there are three forces in the horizontal direction…

In real life, dams typically have a cross-section that is thicker at the base. Also, dams are held in place horizontally by the sidewalls of rock as well as the bedrock.

14.9 Overhead views of two dams are shown at right. Both dams are equally tall. The first dam holds back water in a narrow canyon. The second dam holds back water in a canyon that opens up to a wide reservoir. Which dam needs to be made thicker to prevent collapse? Explain for credit.

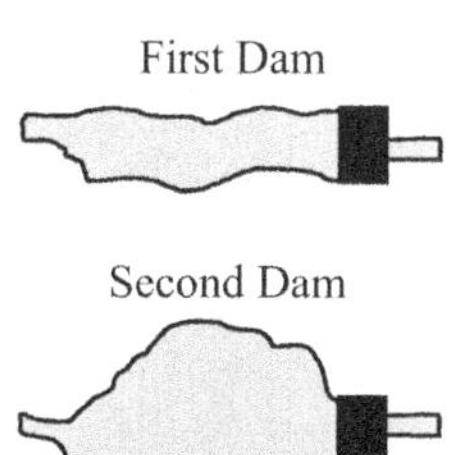

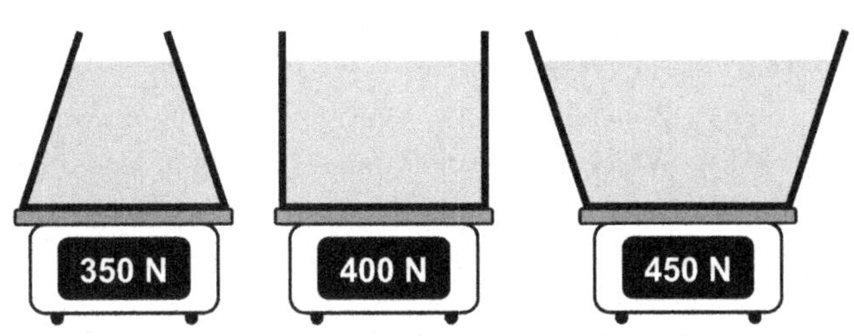

14.10 Three containers hold water on three scales as shown in the figure. All containers have identical bases with equal areas and the containers are filled with water to equal depths. The containers themselves are roughly equal mass. Furthermore, the mass of the containers is almost completely negligible compared to the water inside. The sidewalls of each container are different as shown in the figure. Obviously, the container with the most water exerts the largest force and the most pressure on the scale. This appears to contradict the equation that says water should have equal pressure at equal depths. Determine an explanation resolving this apparent contradiction.

The funky glass tube filled with water

14.11 A common demonstration apparatus for physics classes is shown at right. Notice the depth of water in each tube, regardless of the funky shapes, is equal. This implies the pressure is equal at the bottom of each tube.

Suppose, now, you pour additional water into the leftmost tube. For a brief instant the leftmost tube would have a slightly higher water column. How would this affect the forces acting on the small chunk of water indicated by the dark grey cylinder?

Eventually the fluid in the tubes will reach equilibrium again. The depths of the water columns in each tube will again be equal. You may have heard the saying, "water seeks its own level". This brief discussion is my version of thinking about that statement.

Popping the bottom out of a bottle

Pascal's principle states that when the pressure increases at any point in a confined fluid there is an equal increase at every other point in the fluid. A neat way to demonstrate Pascal's principle is to fill a glass bottle with water. Then put a rubber stopper in the opening of the bottle. Hit the rubber stopper with a mallet. Applying the large force on the small area stopper causes a large pressure increase at the top of the bottle. This pressure increase at the top of the bottle ultimately results in an identical pressure increase at the bottom of the bottle despite the larger area. Boom!

Pascal's Blaising Barrel

There is a really cool video if you do a web search for "Pascal's Blaising Barrel". A cool related demo can be found by doing a web search for "inverted Pascal experiment".

14.12 Hydraulic lift

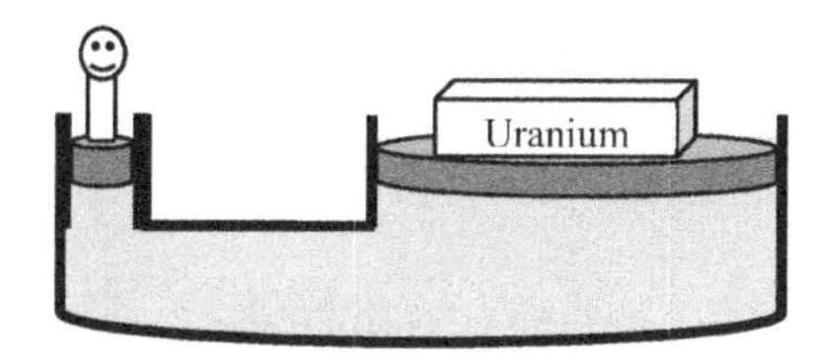

Hydraulic systems make practical use of Pascal's principle. The figure at right shows a hydraulic lift. Esmerelda stands on a piston on the left side of the lift with mass 45 kg (about 100 lbs). Under that piston is a reservoir filled with an incompressible fluid. On the right side of the lift is a large block of uranium. Don't worry, its depleted so Esmerelda's cool with it. Esmy's piston has a diameter of 20.0 cm while the uranium rests on a piston with a diameter of 63.2 cm. At this instant in time, the depths of the fluid column on each side of the lift are roughly equal and the system is in equilibrium. Note: near earth's surface 1 kg $\approx$ 2.2 lbs. Last comment: assume the masses of the actual pistons, the dark grey cylinders supporting Esmy and the Uranium, have negligible mass.

a) Sketch a free body diagram for Esmy's piston. Use this diagram to determine the pressure at the top of the fluid column just below Esmy's piston.
b) Sketch a free body diagram for the other piston and determine the weight of the depleted uranium used to keep the system in equilibrium.

14.13 Hydraulic jack

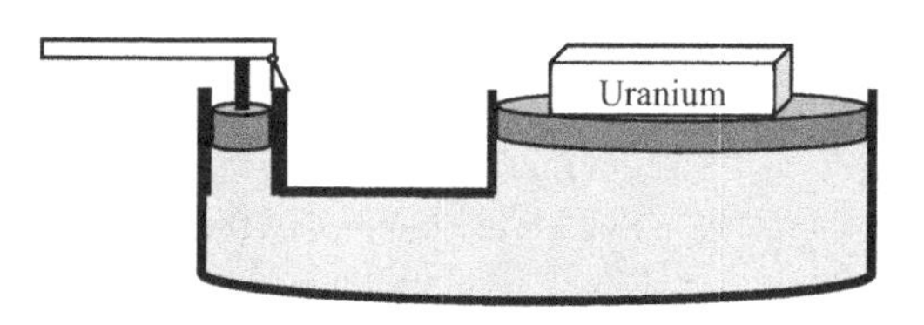

This is a variation of the previous problem. The left and right pistons have diameters of d and D respectively. The uranium has a weight of w. A fulcrum is placed on the edge of the left piston as shown. Assume the rod connected to the fulcrum has length $L = 6d$ and negligible mass. The thick black line represents a stiff piece of metal which connects the center of the piston to the rod. At this instant in time, the depths of the fluid column on each side of the lift are roughly equal. A person exerts a force on the left end of the rod to keep the system in equilibrium.

a) Determine the upward force exerted on the rod by the stiff piece of metal. Answer in terms of w, d, and D.
b) Determine the force exerted by the person to keep the rod in static equilibrium. Answer as a simplified fraction times w.
c) Assume $D = 3.16d$ and the human can comfortably exert a force of 50 lbs downwards with her hand. What weight of uranium would they be supporting? Answer in tons.

The point of this problem is to show how a person can use both lever arms and hydraulics in tandem to hold up objects much heavier than they are. While this problem is somewhat simplified/unrealistic, it gives you an idea of how humans can use jacks to raise and lower an automobile.

Try a web search for hydraulic brakes at this moment.
It is pretty amazing to consider how a tiny force from a foot can cause a giant bus to slow down!

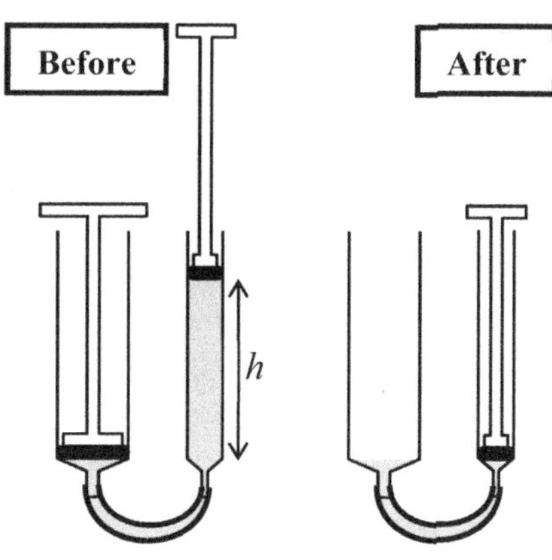

14.14 Force distance trade-off for hydraulics

In the previous problems we see that a small mass can hold up a much larger mass. In many practical applications we need not only to hold the heavier object in place but also to raise/lower it.

Consider the simplified version of a hydraulic lift constructed from two syringes shown at right. The syringes are essentially cylinders and the grey material inside is water. The right syringe has diameter d while the left syringe has diameter $2d$. I connected the nozzles of each syringe with some tubing and some zip ties to prevent leaks.

Notice in the before picture the plunger of the skinny syringe is at full extension. In the after picture it is compressed all the way. **Note: I did not draw the other plunger because you will be figuring out where it goes!**

a) Determine the volume of water initially inside the cylindrical portion the skinny cylinder. Answer in terms of d and h.
b) Upon compressing the skinny cylinder, that volume of water must flow into the larger syringe. Determine the final height of the water in the larger syringe. Answer as a fraction times h.
c) Sketch the plunger in the appropriate location for the AFTER figure.

If a force of F is applied to the plunger on the skinny syringe, we expect this to cause a force of $4F$ to be exerted by the other plunger. Conversely, if the skinny plunger is pushed a distance x the larger plunger only moves distance $x/4$. Notice that the work corresponding to each plunger's motion has magnitude $W = F$!

If you are interested, this would be a great time to read about how hydraulic brakes and power brakes work. You should now be able to understand what you read about them online. There are numerous excellent resources. Try a web search for "hydraulic disc brake diagram".

Absolute pressure versus gauge pressure

When you measure the pressure in your tire or an inflated ball you use a gauge. Perhaps you have seen a compressed gas cylinder used for filling balloons that has a gauge to show how much air is left in the tank. Vacuum chambers such as bell jars will often have a gauge to register the pressure of the air inside. It is important to mention what pressure is indicated by the gauge.

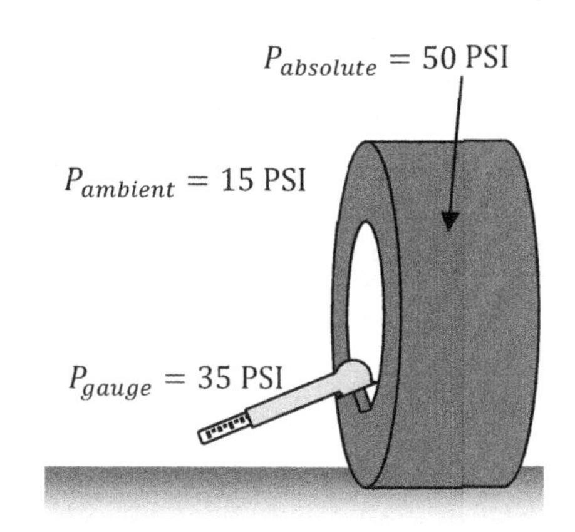

Suppose you read a tire pressure gauge at 35 PSI. The pressure of the air inside the tires is **not** 35 PSI! The gauge measures pressure differences. The gauge pressure is 35 PSI but the actual pressure (called absolute pressure) inside the tire is 35 PSI greater than ambient pressure. The pressures are related by the equation

$$P_{gauge} = P_{absolute} - P_{ambient}$$

Perhaps you are familiar with the equation $PV = nRT$ from chemistry. It is worth noting that the pressure used in this equation is $P_{absolute}$ as opposed to the pressure measured in a lab experiment typically given by P_{gauge}. However, for a great many situations, gauge pressure can be more convenient for computations.

I had an idea for a demo I have yet to make...

Get a tall water column...at least 1 m tall, 2m would be ideal. The diameter must be big enough to insert a glass. An 8-ft acrylic pipe with maybe 4-inch diameter would be perfect but probably expensive...4-ft is ample and less unwieldy. Weld on an end cap to the pipe so it holds water. And fill it most of the way up.

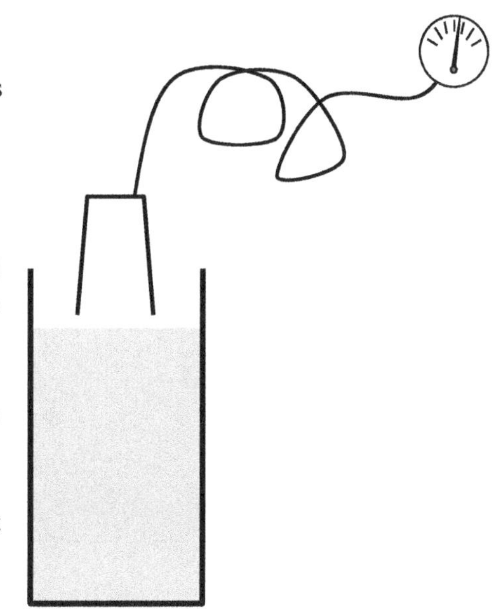

Next, we want to invert a glass jar or plastic equivalent and push it into the water upside down. The goal is to learn about the air bubble trapped inside the inverted glass and you push it deeper into the water. I want to think about both the volume and pressure of the trapped air bubble.

To quantify the pressure, I thought it would be nice to somehow attach a pressure gauge to one end of a hose. The other end of the hose would connect to the inverted glass underwater. In addition, to make it easier to push the glass beneath the surface of the water, one could attach a 3-ft dowel rod, a couple of metersticks, a pipe, or something similar.

a) As the cup is pushed beneath the surface of the water, how what is the absolute pressure of the air inside the inverted glass?
b) What is the pressure displayed by the gauge?
c) How far must one push the glass below the surface for the gauge pressure to be 0.5, 1.0, and/or 2.0 atm?
d) Going further: if you know $PV = nRT$ and we assume air is an ideal gas we can also discuss the predicted volumes of the bubble. Assume the air temperature inside the bubble is essentially constant. If the inverted cup is placed 1 meter below the surface of the water, we know the absolute pressure inside the cup is 2 atm (twice normal atmospheric pressure). This means the air pressure had doubled from the time it was at the surface. How should the volume of air in the inverted cup change?

Buoyant force (Demo, similar to 14.19 with mass balance under beaker, scales always sum to constant)
The simplest way to describe the buoyant force is to know this statement:

The buoyant force on an object has magnitude equal to the weight of the fluid displaced.

While this definition is excellent for a conceptual understanding, it is nice to have an equation as well. The equation for a buoyant force B is given by

$$B = \rho_f V_{disp} g$$

where is the ρ_f density of the fluid (not the object), V_{disp} is the volume displaced by the object experiencing the buoyant force, and g is the magnitude of the acceleration due to gravity. There several points worth emphasizing:

- ρ_f density of the fluid (not the object)
- V_{disp} IS the volume of the object when the object is solid AND fully submerged
- V_{disp} IS NOT the volume of the object when the object is hollow
- V_{disp} IS NOT the volume of the object when the object is partially submerged

Buoyant force problems typically involve the weight of the object as well. It is often useful to write the weight of the object as

$$w = \rho_{obj} V_{obj} g$$

instead of using mg. Notice that we are now using the density of the object ρ_{obj} and the volume of the object V_{obj}. There several points worth emphasizing:

- ρ_{obj} density of the object (not the fluid)
- V_{obj} can require some effort for hollow shapes
- V_{obj} is the same whether the object is submerged or not

14.15 Explaining buoyant force using pressure versus depth
Consider a side view of a cylindrical block of height h and radius r.
The top of the block is distance z below the surface of the water.
Several arrows are drawn near the top indicating the pressure (and thus force) the water molecules are exerting on the block.

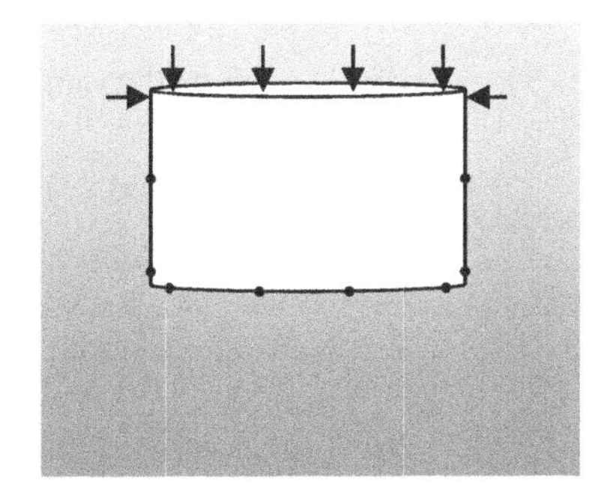

a) What is the pressure difference between the top and bottom of the block?
b) Draw arrows for each of the black dots indicating the size and direction of the water pressure at each black dot. Don't worry about having things perfectly to scale but do try to get the relative sizes of the arrows correct.
c) Since each arrow represents a force, what direction is the net force on the block due to the water pressure varying with depth?
d) The net force due to a pressure differential is given by $F_{net} = (\Delta P)A$. Determine the magnitude of the force on the block due to this pressure differential.

Buoyant force on objects at the bottom of a pool

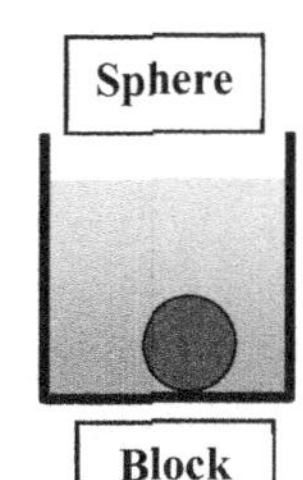

Consider a solid, metal sphere at the bottom of a pool. Most people have no trouble assuming there is a buoyant force up on the sphere. Water is surrounding almost the entire thing so it is easy to see how water could push upwards on the bottom of the sphere. A normal force, a buoyant force, and gravity all act on the sphere at the bottom of the pool!

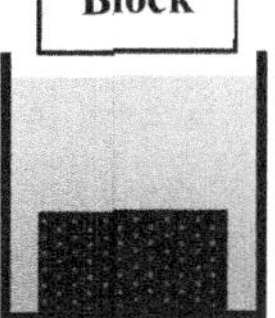

Now consider a metal block as shown in the second figure. A normal force, a buoyant force, and gravity all act on the block at the bottom of the pool! Students have trouble with this as they can't see how the water can push up on the block when it is touching the bottom of the pool. Consider the block as it sinks, it is clearly displacing water and thus has a buoyant force upwards. When the block reaches the bottom of the pool it is still displacing water and thus still has a buoyant force.

Consider a cube of plastic less dense than water sitting at the bottom of an empty pool. At this point the only forces are weight and normal force. The tank is then filled with water. Do you expect the plastic to stay in contact with the bottom of the pool or to float upwards? If you expect it to float up this might help you feel better about believing there is a buoyant force on an object at the bottom of the pool. Check the density of a hockey puck or maybe some thick foam core…If hockey puck is less dense you *might* be able to get some contradictory results if the surfaces are properly prepared? I'm not sure but I want to try it out. I'm talking to myself here I guess…

Just barely floating

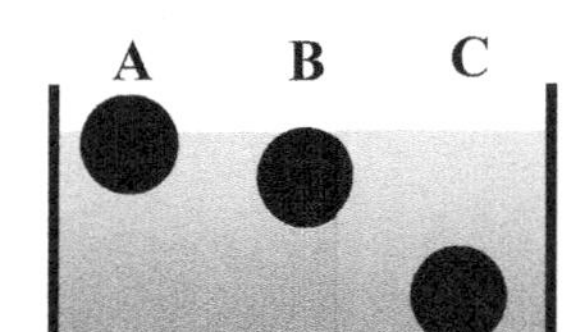

14.17 Three bowling balls are placed in a pool of water. Each ball has diameter 217 mm. Ball **A** weighs 10.0 lbs while ball **C** weighs 14 lbs. The weight of ball **B** is unknown. Ball **B** just barely float (floats but is also fully submerged). Note: near earth's surface 1 kg corresponds to 2.2 lbs.

a) Rank the buoyant forces exerted on the balls.
b) Rank the densities of the balls.
c) Sketch a free body diagram and determine a force equation for ball **B**.
d) Determine the weight of ball **B** in lbs.

Floating object partially submerged

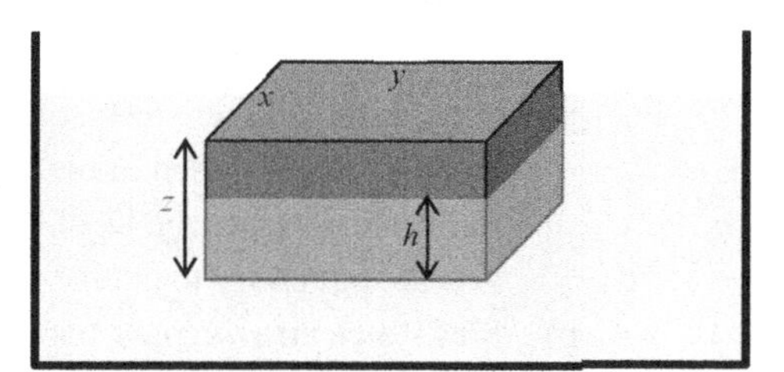

14.18 A rectangular solid with length x, width y, and height z has density ρ. It floats partially submerged in a fluid of density ρ_f.

a) Is the buoyant force greater than, less than, or equal to the weight of the object?
b) Is the volume of fluid displaced greater than, less than, or equal to the volume of the object?
c) Draw a free body diagram and determine the force equation.
d) Determine the fraction of the object's height that is below the surface of water in terms of the densities. Said another way, I expect $h = fz = \frac{?}{?}z$ and I want you to determine the fraction f.

Hydrostatic weighing

14.19 The goal of this procedure is to determine the density of awkwardly shaped objects. A solid object hangs from a spring scale and its weight is recorded as w_1. The object is then lowered into a pool of fluid with density ρ_f and the scale registers an underwater "weight" w_2. Note: scales read *tension*, not *weight.* Non-physicists say the scale reads weight. I phrased this problem to match how a lay person might say it. Note: since the object remains fully submerged, we know its density must be greater than the fluid.

a) Draw the FBD and determine the force equation for the submerged object.
b) Determine the density of the object in terms of $w_1, w_2, \& \rho_f$.
c) Suppose one repeated this experiment with a less dense fluid. Would you expect the experimentally obtained value of the object's density to have more or less uncertainty?
d) If you wanted more and more sig figs on your measurement, perhaps you would try to increase the density of the fluid more and more. At some point, increasing the fluid density to increase the number of sig figs would no longer be effective. What is the upper limit on the fluid density to use this procedure? What are the scale readings in that limit?

Note: hydrostatic weighing is often used to determine a person's body fat percentage. The underwater "weight" a *person* hangs from the scale (or sits on a scale at the bottom of the tank).

14.20 In the past, boats have been made from concrete. One example is the SS Palo Alto in Aptos, CA.
A boat is made from concrete with density 2400 kg/m^3. Notice this is actually slightly less dense than aluminum (a common canoe material). The simplest design I could think of was a boat with uniform thickness t and exterior dimensions x, y, and z as shown. To be clear, the thickness of the bottom of the boat is also t.

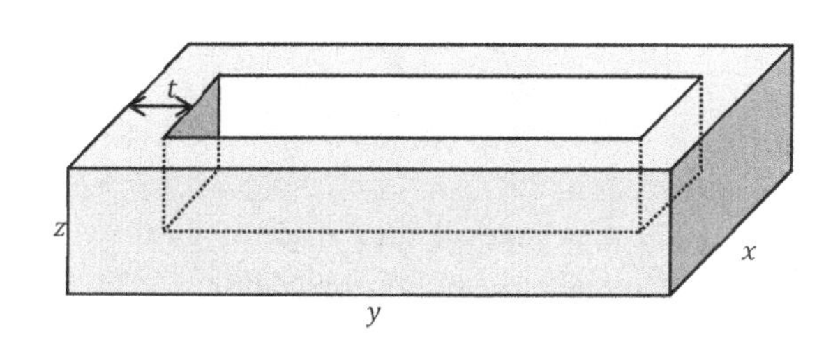

a) Determine the volume of concrete used to make the boat in terms of x, y, z, and t.
b) Suppose the boat is 6.00 m long, 1.00 m wide, and 14.3 cm thick. What minimum depth is required for the boat to float? Tip: for this problem plug in numbers right away to spare the pain.
c) **Challenge:** Suppose the boat is 6.00 m long, 0.800 m wide, and 1.25 m tall. What is the maximum allowable thickness of concrete to be used if the boat is able to float? Again, I plugged in numbers right away. Note: any useful boat requires a smaller thickness to handle passengers and/or cargo.

Note: in practice ferrocement is typically used. Ferrocement employs iron mesh, rebar, or something similar to improve the structural characteristics of the boat. Check out the Concrete Canoe National Competition put on by the American Society for Civil Engineers if you are interested in learning more. Cal Poly has frequently competed and has won several times.

14.20½ An aluminum pipe has density ρ_{Al}, height h, inner radius $\frac{2}{3}R$ & outer radius R.
The pipe is suspended from a light spring while halfway submerged in a fluid of density ρ_f.
In equilibrium, the spring is stretched distance y.

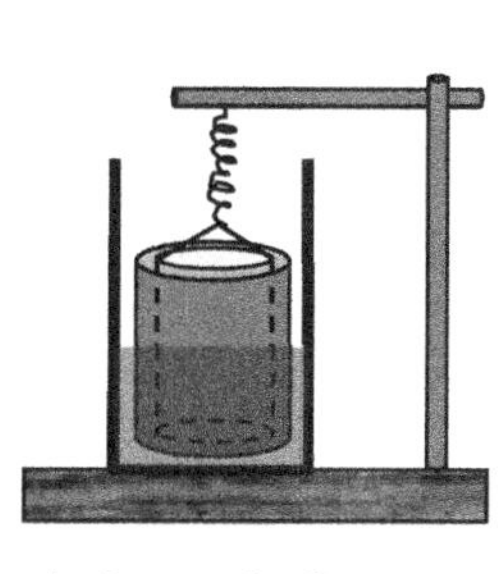

a) Determine an algebraic expression for the spring constant in terms of the givens and g.
b) How does the work for part a change if the hole at the bottom of the cylinder is sealed with waterproof tape *before it is placed in the fluid*? The tape has negligible volume and mass. Assume the spring stretched the same amount.
c) How does the work from part a (pipe with no tape) change if the spring is *compressed* instead of *stretched*?

If you desire practice on a hollow shape with <u>spherical</u> geometry, try 14.42½...

Center of Buoyancy & Stability
The center of buoyancy is defined as the center of the displaced volume of fluid. You could imagine making the submerged portion of a boat a solid, uniform density object and find its center of mass.

To be stable while floating one typically assumes the center of mass must lie below the center of buoyancy. I find it interesting that in some cases the center of mass can be above the center of buoyancy and yet the object can still be stable! While this is not a topic for a freshman physics course, if you are interested in the stability of boats you can probably read about it online now and make some sense of it if you are patient. You will learn about topics such as the metacenter and righting arm. Cool stuff!

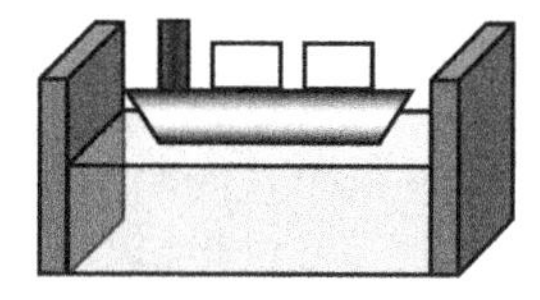

14.21 A lock is a device used for raising and lowering ships in a canal. For our purposes, it is essentially the same thing as a swimming pool big enough to hold a boat. A ship loaded with iron rods and low-density solid, plastic rods is in equilibrium in a lock. Captain Vega is in charge of running an unusual physics experiment. She orders the crew to dump all the iron into the lock.

a) What happens to the water level in the lock? Does it go up, down, or remain the same? Explain why.
b) How is the answer to the previous question affected if the crew had instead thrown the low density plastic rods overboard? Explain why.
c) Challenge: determine what factors will affect the height of the deck of the boat relative to the bottom of the lock for the case of iron being dumped overboard. Determine an algebraic expression including these factors which determines the change in height of the deck. Once you figure it out, check your work using the Phet buoyancy simulation. Set the simulation to two block mode and set the brick on top of the wood.

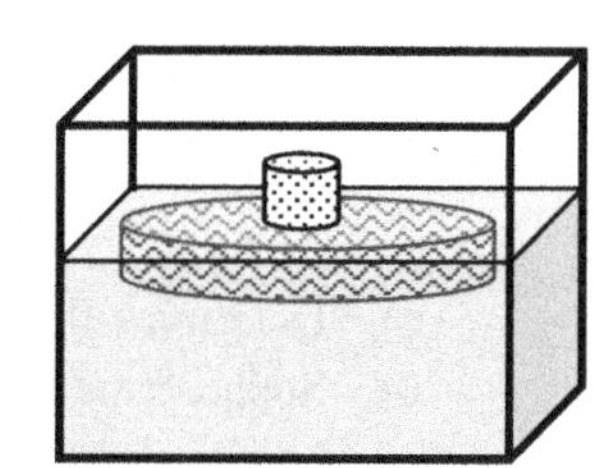

14.22 A cylindrical raft is made of plastic with density 0.033 g/cm^3. The raft has a small cylindrical piece of iron with density 7.87 g/cm^3 affixed to the center of the top of the raft. The raft is being tested in a pool as shown in the figure. Note: approximately 1.5 kg of iron will have the plastic raft (but none of the iron) fully submerged.

a) Determine the minimum volume ratio (plastic volume to metal volume) required to keep the metal above the surface of the water? Let's assume, for now, we can design the system in some stable way.
b) Assume we use a raft with the volume ratio determined in part a. The raft flips over and the metal is now submerged (but still affixed) underneath the raft. What happens to the water level of the pool? Explain.
c) Determine the fraction of raft volume *above* the surface of the water after the raft has flipped.

14.23 A gold sphere, an aluminum block, a scrap of treated lumber, a bag filled with water, a chunk of Styrofoam, and an inflated beach ball all walk into a bar. All objects have the same mass. The objects are then dropped into a swimming pool. Note: you may assume the bag used to hold the water is made of thin, leak-proof plastic with negligible mass and volume.

a) Which objects will be fully submerged? Which are floating? Is it possible to float and be fully submerged?
b) Rank the buoyant forces acting on the *submerged* objects. Alternatively, which has smallest buoyant force?
c) Rank the buoyant forces acting on the *floating* objects. Alternatively, which has the largest buoyant force?

The atmosphere as a compressible fluid

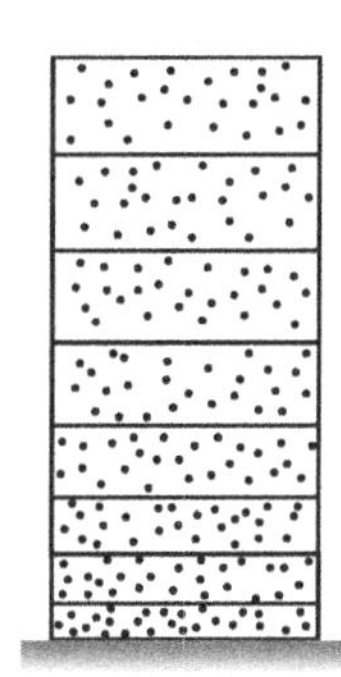

As one travels upwards through the atmosphere, the density of air changes. A fluid which has non-uniform density is said to be *compressible*. A good way to visualize the atmosphere as a compressible fluid is to consider a tower made of stacked moist sponges. The sponges near the bottom of the stack are squished the most while the sponges at the top are almost full size. Since each sponge has the same mass, we see the density of each sponge decreases as you go towards the top of the stack. This is analogous to the density of air increasing with altitude.

You may also be familiar with the fact that temperature changes often occur with changes in air pressure. In a later course we will describe the applicability of the equation $PV = nRT$. A higher level thermodynamics course is often where you learn about the mathematical modelling of pressure and density versus depth for air.

For buoyant force problems with balloons in air, you might be concerned. We derived the buoyant force using an example of incompressible fluid. Fortunately, the key principle of the buoyant force is still valid in air: the buoyant force is proportional to the weight of the displaced fluid. At any arbitrary height in the atmosphere, the density of air doesn't change much between the top and bottom of even large objects such as hot air balloons. Therefore, at any particular height one can use the buoyant force assuming one uses the density of air appropriate for that height in the atmosphere.

14.24 A 69 kg swimmer with a body volume of 70.0L is ready to dive in a pool. Note: 1 cm^3 = 1 mL. The density of the water is 1.00 g/cm^3. By comparison, the density of air is, very roughly, about 1 kg/m^3. Depending on humidity the value is typically closer to 1.2 to 1.3 kg/m^3. For now, assume the atmosphere is 1.25 kg/m^3.

a) Just *after* the dive the swimmer is completely submerged in the water. Is there a buoyant force on the swimmer?
b) Just *before* the dive is there a buoyant force on the swimmer?
c) Determine the ratio of the buoyant force *before* the dive to the buoyant force *after* the dive.

14.24½ A balloon has a 9.0" diameter and is essentially spherical. The balloon itself has mass 2.4 g. To be clear, this is the mass of the latex used to make the balloon with no air inside it. The balloon is inflated with helium (He) of density 0.18 g/L. A 2.00 m long string is tied to the balloon. When released from rest near the ground, the balloon quickly reaches equilibrium with approximately 35 cm of string still touching the ground. Assume the atmosphere has density 1.25 kg/m^3. Figure not to scale.

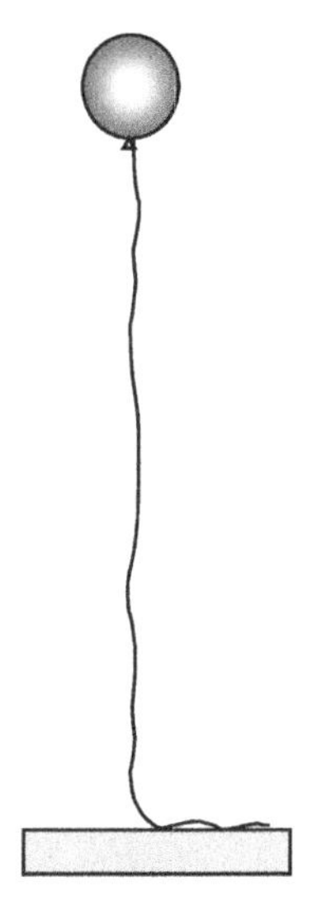

Note: when inflated the balloon compresses the helium (and thus increases the density). Let us ignore this compression and related density change for now.

a) Is it reasonable to ignore any buoyant force acting on the string (when considering forces on the balloon and string)? Explain why or why not.
b) Determine the linear density of the string (in g/m). Try to do it algebraically and plug in numbers at the last step.

Flowing Fluids: Definitions, Assumptions, and Equations

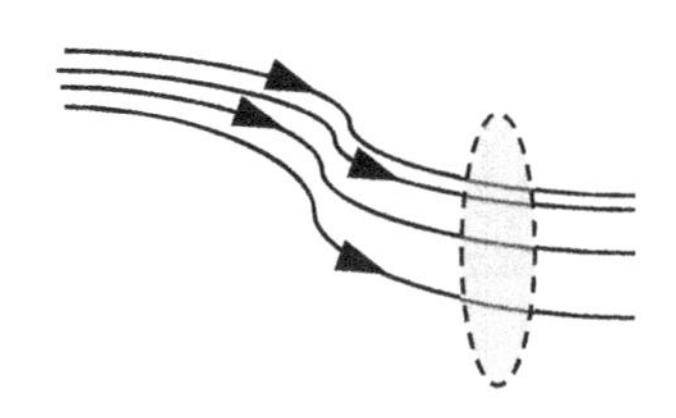

ρ = density of fluid
A = cross-sectional area of fluid flow (grey circle at right)
R = Volume flow rate = flow rate = $\frac{\text{gallons}}{\text{min}}$ or $\frac{\text{Liters}}{\text{sec}}$ passing through a given area
v = fluid speed = speed = $\frac{\text{m}}{\text{s}}$ a chunk of fluid is traveling at a given point
P = static pressure = pressure = force per unit area exerted by fluid perpendicular to flow
$q = \frac{1}{2}\rho v^2$ = dynamic pressure = kinetic energy per unit volume, relates to pressure to stagnation pressure
$P + q$ = stagnation pressure = total pressure of a moving flow if the flow is forced to a stop
Streamline = path a fluid particle travels, streamlines can meet at a stopping point but they will never cross
Laminar flow = no turbulence, smooth streamlines, no vortices

We will focus on the simplest possible fluids in motion. We will idealize these simple fluid flows as

- Non-viscous (no drag)
- Incompressible (constant density)
- Laminar flow (no turbulence, eddies, or vortices)

Warning: very few fluid flows in real life will are model well with these assumptions. That said, we ignored air resistance for everything in chapter 4 and managed to learn a lot about projectiles…right? We gotta start someplace…might as well be as simple as possible. If you get into fluids and want to do aerospace engineering or something like that you can learn about vorticity, Navier-Stokes, and all that cool stuff later.

Eqt'n Name	Equation	Comments
Continuity Eqt'n	$A_1 v_1 = A_2 v_2$	• Fluid in = fluid out for a closed pipe system. • If this wasn't true fluid is leaking out of the system or pressure is going to build up until an explosion occurs.
Bernoulli Eqt'n	$P_1 + \frac{1}{2}\rho v_1^2 + \rho g h_1 = P_2 + \frac{1}{2}\rho v_2^2 + \rho g h_2$ $P + \frac{1}{2}\rho v^2 + \rho g h = \text{constant}$	• Requires incompressible (constant ρ) and no friction • Conservation of energy for two points on streamline • h = HEIGHT (not depth as in pressure vs. depth) • P = pressure = STATIC pressure pointing $\perp$ to flow

Demos:
Benoulli blower, Coanda ball, Flettner flyer, reversible mixing, roll pipe into tank, balloon of water, mesh on jar, three holes in side of pipe, siphon, bed of nails, two balloons, homemade pressure gauge, newspaper chop (× 2), big spoon and aquarium pump, jumbo perfume bottle

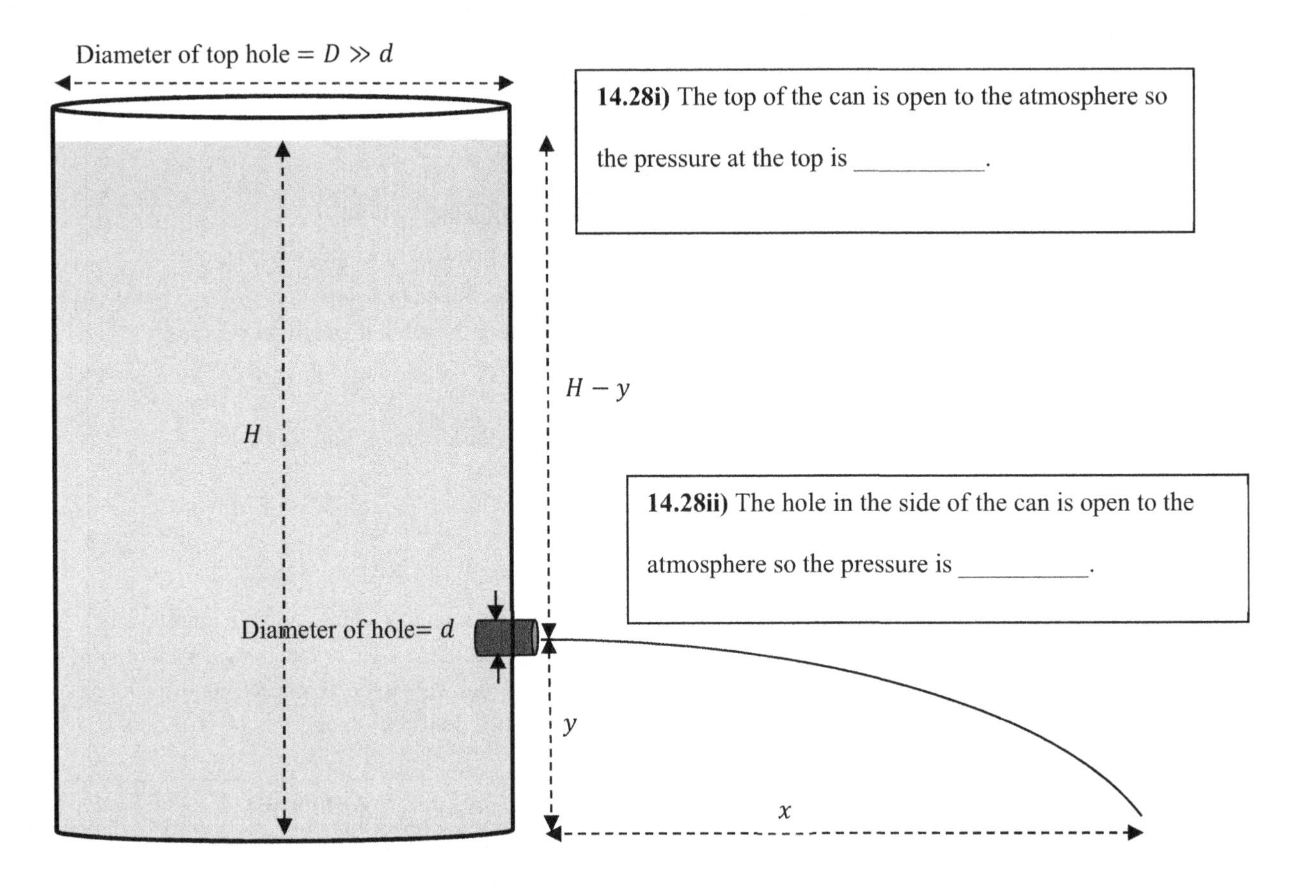

14.28 A cylinder of diameter D is open at the top to the atmosphere. A small hole of diameter d allows the fluid in the can to flow out.

a) Determine the ratio of fluid speed at the top to fluid speed at the hole using the continuity eqt'n.
b) Often $v_{top} \approx 0$. Assuming this, determine the speed that water leaves the hole in terms of H, y, and g.
c) Determine the horizontal distance traveled by fluid leaving through the hole in terms of H & y.
d) **Calc People:** For a given cylinder (for a fixed H), determine the height y which gives the fluid a maximum range x. Compare the range for the following three values of y: $H/4$, $H/2$, and $3/4H$.
e) Under what circumstances is it appropriate to say that $v_{top} \approx 0$?
f) How would this problem differ if the hole was angled upwards by angle θ instead of being perfectly horizontal? What is the max height reached by the fluid relative to the ground? Think: should it be greater than, less than, or equal to H...now try it and see!
g) Think of the special case where the fluid is angled straight up instead of straight out. How far up will the fluid fly? Think: should it be greater than, less than, or equal to H...now try it and see!
h) Now suppose that instead of being open to the air, the top is a sealed and maintained at a pressure P_1. Still assuming that $v_{top} \approx 0$, solve for the speed of the fluid coming out of the hole in terms of atmospheric pressure P_0, P_1, g, H, and y.

Going Further: What about a case where the top diameter is exactly twice as big as the hole diameter? For this case assume the top is open to air as in the original case. Try setting it up but maybe don't bother finishing the whole thing.

14.29 An ornamental fountain is designed with five nozzles shooting water vertically into the air (figure not to scale). The maximum height of each water stream is h = 3.00 m. The diameter of each nozzle is 1.00 cm. The nozzles are flush with the ground which is approximately z = 1.25 m above the centerline of the water main below. The water main has a diameter of 10.0 cm.

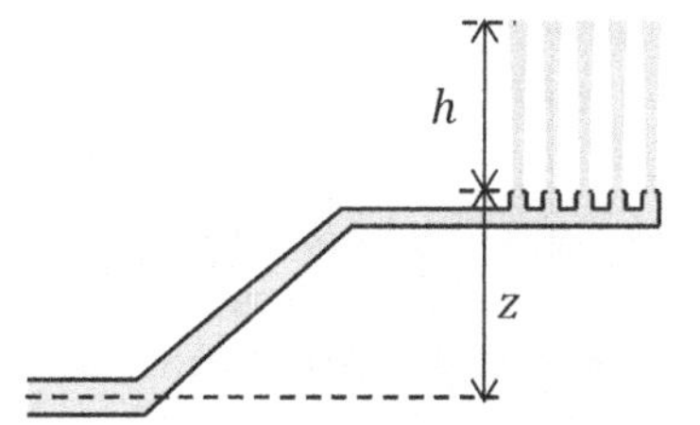

a) Determine the desired speed at which water exits the nozzles.
b) Determine the flow rate required in the mains.
c) Determine the fluid speed in the mains.
d) Is it reasonable to consider the fluid in the mains as being at rest relative to the water exiting the nozzles?
e) Determine the pressure in the mains. Answer in PSI.
f) Suppose one of the nozzles becomes clogged while the flow rate from the main is unchanged. By what factor does the height of the remaining water streams change? By what factor does the pressure in the main change?

14.30 A water siphon is created by running some tubing between two identical buckets as shown at right. Standard atmospheric conditions apply. The diameter of each bucket is very large compared to the tubing diameter. Points **B** and **D** are just inside the openings of the tubing.

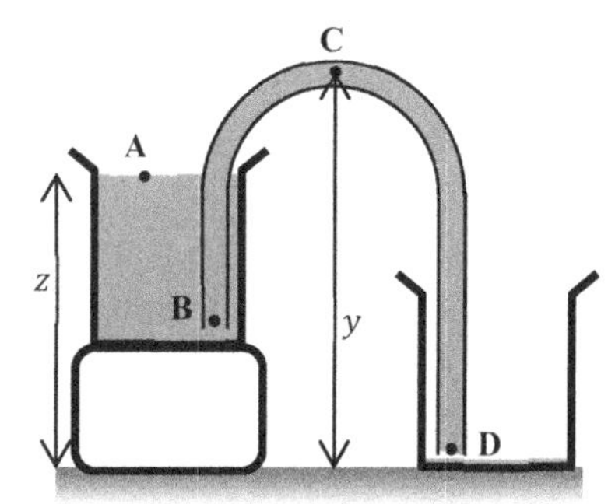

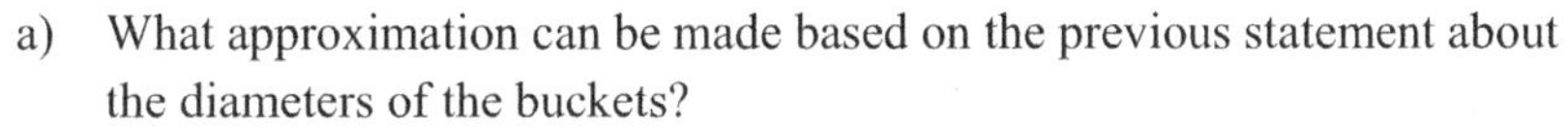

a) What approximation can be made based on the previous statement about the diameters of the buckets?

b) What is the water pressure at points **A** and **D**?
c) If the approximation from the first part of the question was invalid, would it affect your answer to the second part of the question?

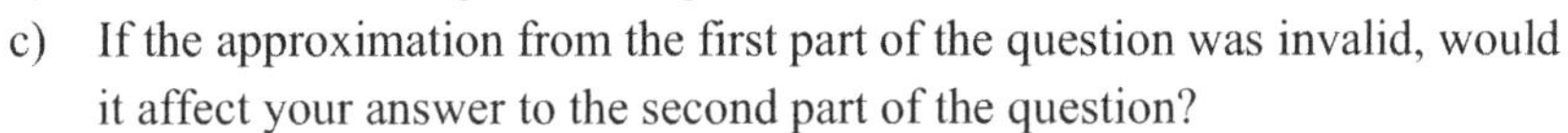

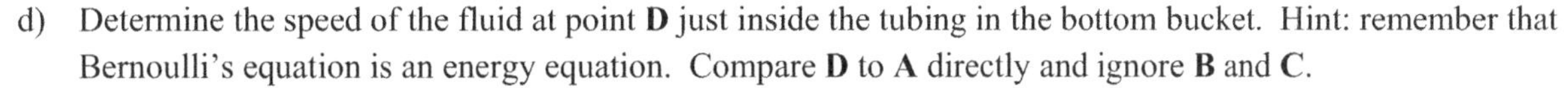

d) Determine the speed of the fluid at point **D** just inside the tubing in the bottom bucket. Hint: remember that Bernoulli's equation is an energy equation. Compare **D** to **A** directly and ignore **B** and **C**.
e) Determine the fluid speed of at points **B** and **C**? Hint: consider the continuity equation.
f) Determine the pressure at point **C**.
g) From a practical standpoint, most siphons will not work if point **C** is close to a perfect vacuum. What then, is the practical upper limit of the height y for this siphon?
h) As the bucket lower bucket fills, what happens to the speed of the fluid in the tube and the pressure at **C**?
i) As a thought experiment, it is interesting to contemplate what would happen if the diameter of the tubing was tapered such that the area of the tubing at **B** is twice as large as the area at **C** which is twice as large as the area at **A**…

Here are some extra questions I thought were good practice on buoyance force and pressure versus depth for a static fluid.

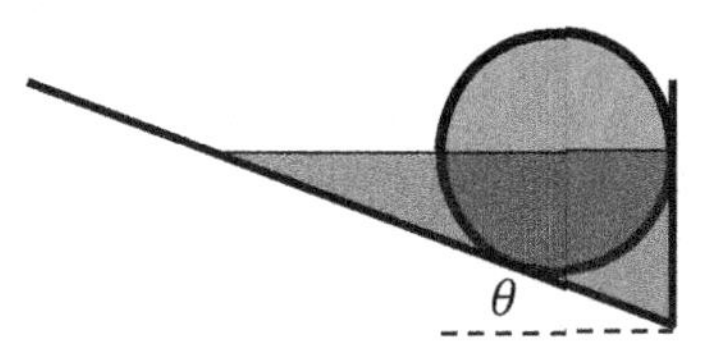

14.30½ A solid sphere of radius R sits at the bottom of a pool. The bottom of the pool has angle θ from the horizontal. The pool is filled with water with density ρ_w until the sphere is exactly halfway submerged. A force sensor embedded in the right wall measures the force between the sphere and the wall as n_1. Friction is negligible between the sphere and all surfaces.

a) Determine the magnitude of the normal force n_2 between the incline and the sphere. Answer in terms of only n_1 and θ.

b) Determine the density of the solid sphere in terms of n_1, R, ρ_w, θ and g.

14.31 When water freezes, it expands. This tells us the density of ice is

A. more than the density of water.
B. less than the density of water.
C. equal to the density of water.
D. none of the above

14.32 A person finds it easier to float in salt water than fresh water. Why?

A. A person weighs less in fresh water.
B. The buoyant force is greater in fresh water.
C. Both A & B.
D. None of the above

14.32½ Explain WHY for your previous answer.

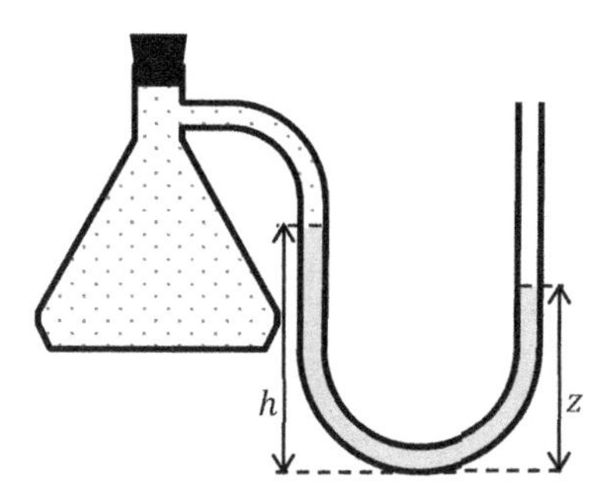

14.33 The flask on the left contains a sealed container of gas from a chemistry experiment. The flask is connected by tubing to a U-tube that contains mercury (shown in grey). The heights of the two sides of the mercury column are shown as h and z. The right side of the tube is open to standard atmospheric pressure P_0. The magnitude of the acceleration due to gravity is g.

a) Determine the *absolute* pressure of the gas in the tube.

b) Determine the *gauge* pressure of the gas in the tube.

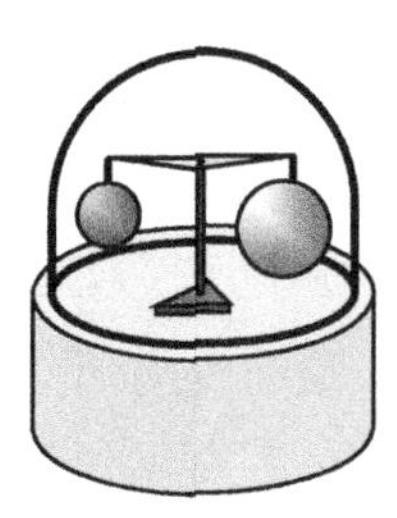

14.34 Two balls in a jar

A hanging balance shows two solid metal spheres of equal mass in equilibrium. One of the spheres is brass with density 8.5 g/cm^3 while the other is aluminum with density 2.7 g/cm^3. The system is currently at standard atmospheric pressure inside the bell jar.

a) Determine the ratio of the radii of each sphere (aluminum to brass).

b) When the bell jar is evacuated, which way (if any) will the balance tip? Explain.

14.34½ One side of a U-tube is filled with a liquid of density ρ. The other half of the tube is topped off with a zombie juice whose density differs by a factor of 5. For some reason we do not know the depth of either column in the U-tube. In hindsight, perhaps drinking some of the zombie juice before doing this experiment was probably not the best idea. We do know, however, the combined depth of the two fluids in the right column is *greater* by distance z as shown.

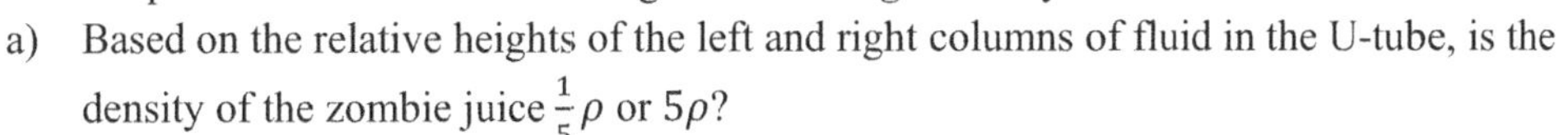

a) Based on the relative heights of the left and right columns of fluid in the U-tube, is the density of the zombie juice $\frac{1}{5}\rho$ or 5ρ?
b) Determine the depth of the zombie juice on the right side.
c) How would your answers change, if at all, if atmospheric pressure dropped by 15%?
d) How would your answers change, if at all, if the experiment was done on Mars with 1/3 of the gravitational force?

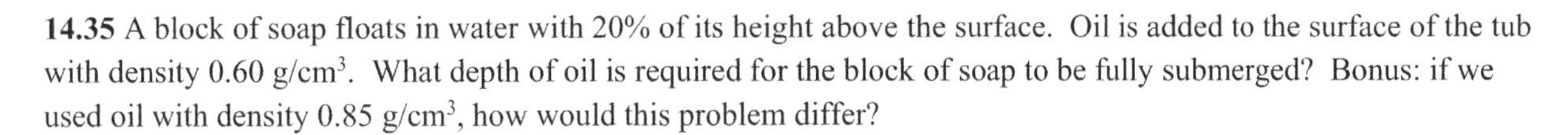

14.35 A block of soap floats in water with 20% of its height above the surface. Oil is added to the surface of the tub with density 0.60 g/cm^3. What depth of oil is required for the block of soap to be fully submerged? Bonus: if we used oil with density 0.85 g/cm^3, how would this problem differ?

14.36 A block floats in two different fluids. In the first fluid half of the block is submerged. In the second fluid 2/3 of the block is submerged. What is the ratio of the densities (first fluid to second fluid)?

14.37 The density of aluminum is $27\underline{0}0\,\frac{\text{kg}}{\text{m}^3}$ while the density of air at standard atmospheric pressure is $1.27\,\frac{\text{kg}}{\text{m}^3}$ on this particular day in lab. A sensitive balance is calibrated and zeroed at normal atmospheric pressure. A second sensitive balance is calibrated and zeroed while inside a bell jar at 0.00001 atm. On each scale rests identical blocks of aluminum. When the bell jar is evacuated to 0.00001 atm its scale reads 9.798 N. Assume the local value of g is $9.798\,\frac{\text{m}}{\text{s}^2}$.

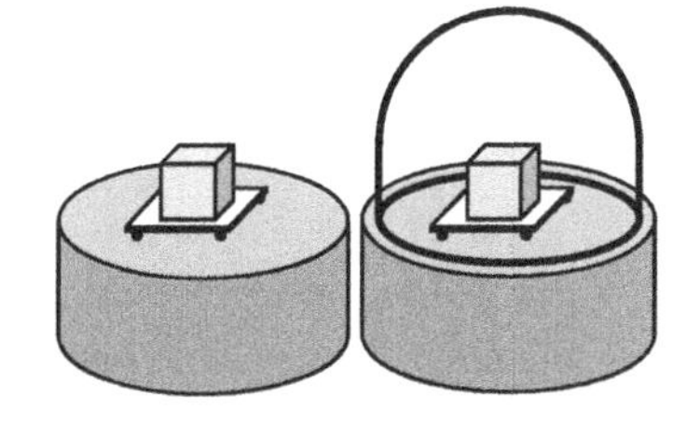

a) Determine the volume of the block of aluminum.
b) Determine the reading on the scale outside of the bell jar.
c) Based on your answer to part b, is it reasonable, in your opinion to neglect the effect of the buoyant force due to air for most scale readings?
d) Which types of materials, high density or low density, would have scale readings more strongly affected by the buoyant force due to air.

Note: Balances this sensitive are affected as much or more by factors such as temperature, air currents, environmental conditions on the day of calibration, where mass is placed, and variation in the local value of g. Believe it or not, it can actually be interesting to read about the inner working of mass balances, sensitive ones in particular.

14.38 Explain how Magdeburg hemispheres are held together despite two people pulling as hard as they can in an attempt to separate them.

14.39 Explain why marshmallows and balloons expand in a bell jar as it is evacuated?

14.40 Explain how a hot air balloon floats in the sky. Why will a hot air balloon eventually reach a maximum height?

14.41 A beaker of water at rest on a balance registers 10.00 N of force. Next, a solid steel sphere is suspended on a string such that exactly half of the sphere is submerged in the water in the beaker. How, if at all, does the scale reading change?

Goes up goes down stays the same

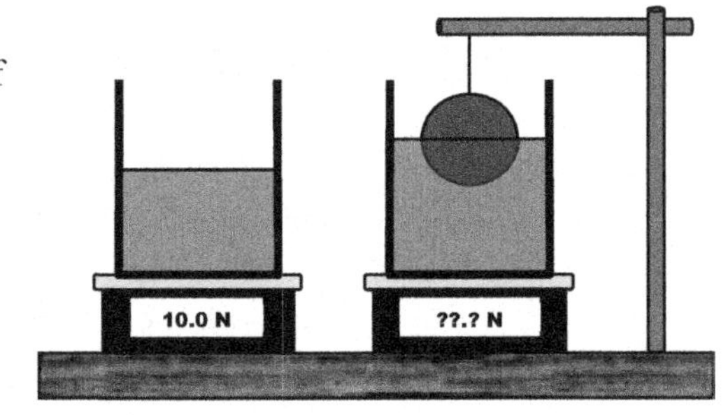

14.42 Suppose the *steel* sphere of the previous problem was replaced by a solid *aluminum* sphere of equal mass. The density of aluminum is about 2.7 times that of water as opposed to the density of steel which is about 8 times that of water. Would the scale reading change by more or less than the previous problem?

More less same amount

Consider a plunger is stuck onto to the bottom of an empty swimming pool. The pool is then filled with water and the plunger is fully submerged. Should the plunger be considered part of the floor of the pool where the water exerts only a downwards force on it. Should the plunger be considered as displacing fluid and have an upwards buoyant force on it? Is this type of problem different than a block at rest on the bottom of a pool? Does it matter if the plunger is only partially submerged or fully submerged? Could you devise an experiment to test your hypothesis?

14.42½ A hemispherical bowl is made from plastic of density $\rho_p = \frac{1}{3}\rho_{water}$. The bowl has outer radius R and inner radius $0.900R$. A mass M is placed in the bowl and is then placed in an empty aquarium. The aquarium is then filled with water up to the edge of the bowl. To be clear, once the tank is filled, the bowl is still resting on the bottom of the aquarium. Determine the normal force acting on the bowl.

14.43 Bed of Nails

As a classroom demonstration, a student is asked to lie down on a bed of nails. A second bed of nails is placed on top of her or his chest. On top of that second bed of nails a cinder block is carefully balanced. Using an 8-lbs. sledgehammer, a trained individual smashes the cinder block with the sledgehammer. Do not try this at home or without an experienced bed of nails expert.

a) Our nails each have a point sixe of approximately 1 mm square. The nails are laid out roughly on a 1" square grid. Suppose the surface area of the average human body (including front, back, sides, arms, etc) is about 1.8 m^2. Estimate the surface area in contact with the nails to determine the number of nails actually being touched.
b) Assuming the average person is roughly 70 kg, determine average pressure at the tip of each nail as it presses into the flesh.
c) During the collision of the sledgehammer with the block, the block shatters. How does this reduce the risk of injury? Discuss factors like impulse, average force during the collision, and collision time as well as energy lost to deformation of the block.
d) A variation on this experiment is done by using a large heavy block that does not shatter upon impact (an anvil). Even though a large force is applied to the anvil as it is whacked by the hammer the person is not pierced by the nails. Explain why using anvil reduces the injury risk during the collision. Of course, getting the anvil on the person in the first place seems non-trivial to me…think how bad it would be if you dropped it while trying to place it on the person.

14.44 My Hands Felt Just Like Two Balloons?

Two balloons are inflated to different sizes. The balloons are then connected to the pipe shown in the figure. At the halfway point in the pipe, a closed valve prevents air from being transferred between the balloons. Once the valve is opened, what should happen to the size of each balloon?

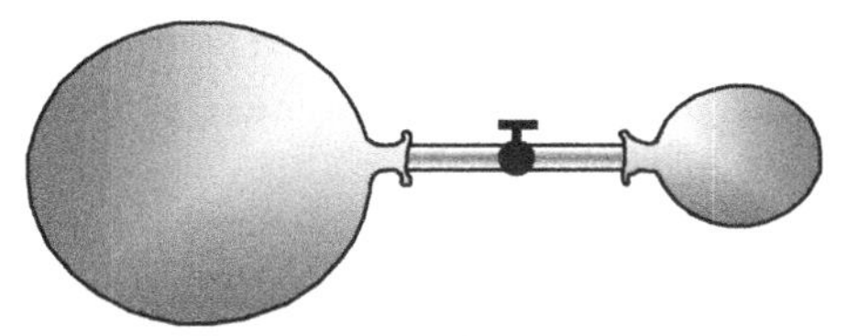

Demo tips: inflate two identical balloons to the same size but do not tie a knot. At this point the balloons are both pre-stressed the same amount. Then let all the air out. It helps if the smaller balloon is a lot smaller. I'm estimating the full one is perhaps ¾ inflated while the smaller one is ¼ inflated (or maybe even less).

14.45 Suction Cup

A suction cup is pressed against the wall. It sticks! Explain how a suction cup works. Challenge: what surfaces work best for suction cups and why? I'm guessing you can figure this out using the internet.

14.46 Drinking with a straw

Explain how a person is able to make fluid move upwards in a drinking straw. What must be happening inside the human body that allows the fluid to go upwards? Are you pulling the fluid up or is it pushed up?

14.47 Newspaper Karate Chop

I obtained a newspaper and several pieces of wood that are about 1" wide by ¼" thick and about 36" long. I set the wood on the table such that a small portion (maybe 6-10") protrudes over the edge of the table as shown. Use appropriate safety precautions if you attempt this demo…

a) What should happen if I give the protruding end of the board a karate chop with my hand?
b) What should happen if I reset the experiment using a single sheet of newspaper carefully prepared to lie as flat as possible over the portion of the stick on the table?
c) What should happen if I repeat the first experiment (no newspaper) but instead of my hand I use a vigorous chop with a broomstick?

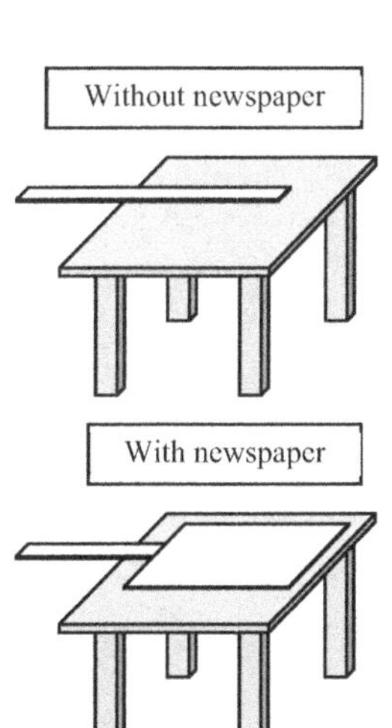

14.48 Make your own sensitive pressure gauge

Get maybe 20 feet of tubing. You might try ¼" inner diameter vinyl tubing. Secure some of it into the shape of a U-tube that is perhaps a meter tall. Leave a bunch of the tubing free for you to move around. You should be able to find plans online if you search for build water manometer or DIY water manometer. Water is much more sensitive than mercury but if you need even more sensitivity you can angle the U-tube. If you do look online you'll see a lot of videos showing people using this kind of thing to sync their carburetors. Evidently physics can be useful!

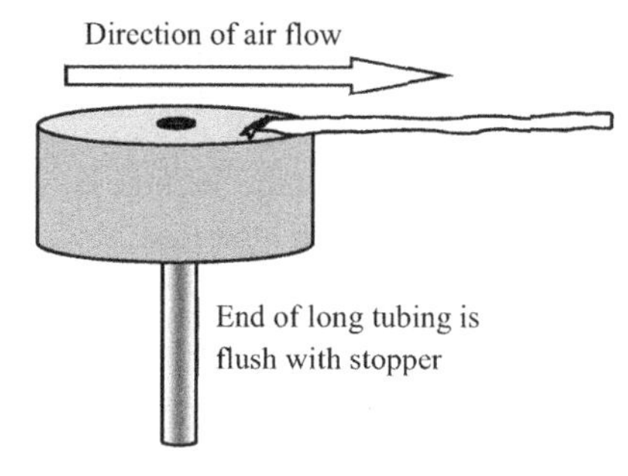

At the end of the long section of tubing I like to place a stopper such that the end of the tubing is flush with the hole in the stopper. I also tape a piece of tinsel to the stopper. The tinsel is a rough guide used to ensure the face the rubber stopper is parallel to the air flow. The stopper can now be placed in any airstream to measure static pressure P_s as long as one ensures the face of the stopper is parallel to the direction of the air flow. Note: without the stopper in place, the end of the tubing will bend the airflow (Coanda effect) and cause erroneous pressure readings. Try putting this pressure sensor inside the airstream on the output port of a large carpet blower and compare the static pressure in the moving airstream to standard atmospheric pressure. The results might surprise you!

Alternatively, you could align this sensor such that air flows straight into the tube (rather than across the face of the stopper). When used in this manner it is called a pitot tube. This moving air is forced to come to rest at the opening of the pressure sensor; it is said to stagnate. This pressure is called the stagnation pressure or pitot pressure. This is the pressure of motionless air at that particular altitude. The stagnation pressure is labeled P_t and sometimes called total pressure. *Total* pressure (P_t) relates to *static* pressure (P_s) by

$$P_t = P_s + \frac{1}{2}\rho v^2$$

a) A pitot tube and static pressure sensor are used in combination simultaneously on an aircraft. Determine the speed of the aircraft in terms of P_s, P_t, and ρ.

b) Suppose the static pressure tube ices up at high altitude. The plane then descends to a lower altitude. Inside the static pressure tube the air pressure is essentially locked at the static pressure of the higher altitude. The pilot computes speed using the formula above but is now using the wrong value of P_s! Is the pilot's calculated speed greater than or less than the true speed of the aircraft?

For more on this do a web search for "pitot static system".

14.49 Reversible fluid mixing on the cheap

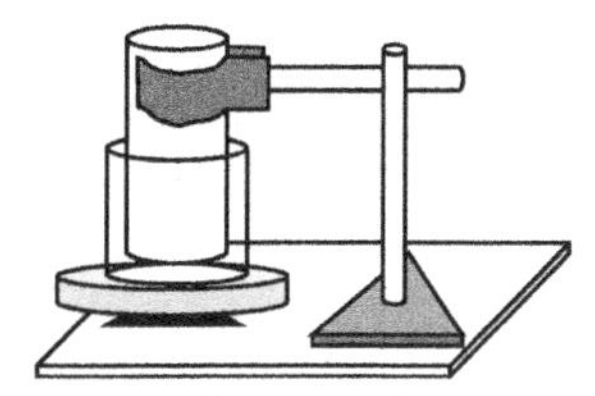

Obtain two beakers or glasses of water with slightly dissimilar sizes. I use a small beaker and a graduated cylinder removed from its stand. Set the two up so the larger is free to rotate on a turntable of some kind while the smaller is fixed and nearly touches the bottom of the other beaker. Fill the gap between the beakers with corn syrup. In a separate beaker, mix in some food coloring to a little bit of corn syrup. Gently inject the colored syrup into the syrup-filled gap between the cylinders. Slowly spin the beaker while counting the number of turns. Ten turns is usually enough to cause the colored corn syrup to mix into the rest of the fluid and make the mixture seem homogenous. Now slowly spin the lower beaker the other way for exactly the same number of turns. Cool.

14.50 Magnus Effect

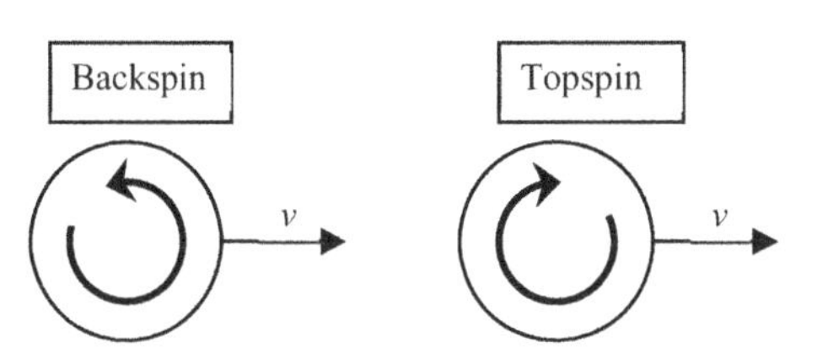

A spinning object travelling through a fluid demonstrates the Magnus effect. I will restrict our discussion to cylinders or spheres with topspin or backspin. A ball has topspin if the top of the ball spins in the same direction as the ball's velocity. Note: from the standpoint of the balls air flows opposite the direction of v. It suffices to say that the rotation of the ball deflects the air as it flows around the ball. Backspin balls deflect the airstream downwards slightly (producing an upward lift force on the ball) while the opposite is true for topspin balls.

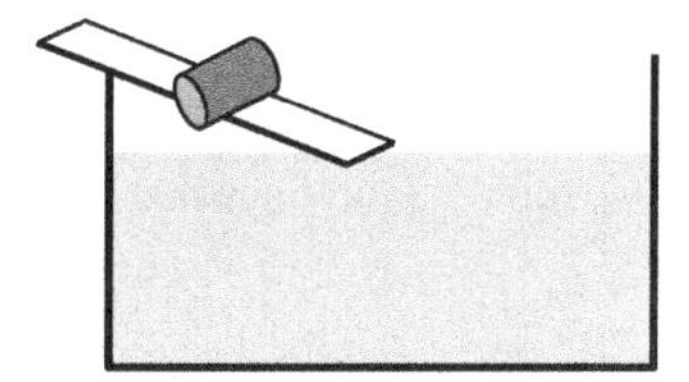

a) Roll a 1" aluminum cylinder down an incline into an aquarium such that the cylinder is spinning as it enters the water near the middle of the aquarium. Should the Magnus effect tend to deflect the cylinder above or below its path towards the bottom right corner of the aquarium?

14.51 Viscosity

So far we have completely ignored any frictional forces for fluid flow in a pipe. Viscosity (η) is typically thought of as the thickness/gooey-ness of fluids. A more technical definition of viscosity might be "the ratio of the tangential frictional force per unit area to the velocity gradient perpendicular to the direction of flow of a liquid".

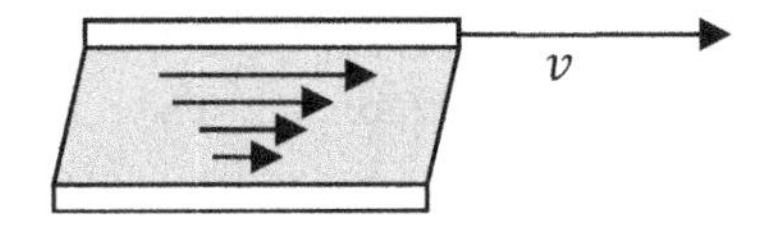

Viscous force can be described by considering two plates with a layer of fluid between them. The bottom plate is stationary. A constant force F (not shown in figure) pulls the top plate to the right with constant speed v. The arrows in the figure represent a velocity gradient. In laminar flow, the speed of the fluid increases from zero at the stationary plate to full speed v at the top plate. The viscous force is given by

$$F = \eta A \frac{\Delta v_x}{\Delta y}$$

where A is the area of the plate in contact with the fluid and $\frac{\Delta v_x}{\Delta y}$ is the velocity gradient. Notice it considers the change in <u>horizontal</u> speed as one goes <u>upwards</u> through the fluid. Side note: in a cylindrical pipe the speed increases from zero at the pipe walls to a maximum in the middle of the pipe.

14.52 Poiseuille's Law (pronounced PWAH-ZUH-EE)

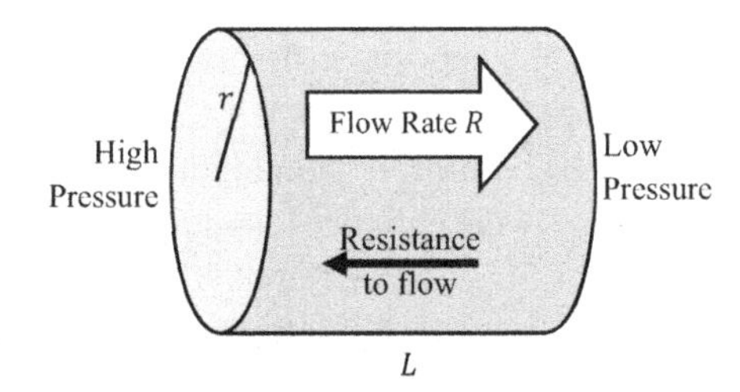

If a viscous force is present in a pipe a pressure difference is required to drive the fluid through the pipe. The pressure difference does work to maintain fluid flow against the negative work of the viscous force. Poiseuille's law states that volume flow rate is given by

$$R = \frac{\pi r^4 \Delta P}{8 \eta L}$$

An interesting application of this comes in blood vessels. In fact, this is what interested Poiseuille and why he studied this in the first place. Notice the strong dependence of the flow rate on the radius. Think about the dramatic change in radius between your aorta (approx. 5 mm) and your capillaries (approx. 0.01 mm) and how this must affect the blood flow rate throughout your body. Thinking about it another way, if your arteries clog by just 19% your blood pressure doubles (to maintain the same blood flow rate).

Lift

Lift is easily the least understood force related to fluids. The following theories of lift are <u>not</u> true:

- NOT TRUE that air flows faster over the top of the wing to meet up with the air in the back of the wing.
- NOT TRUE that lift is generated solely by air hitting the bottom of the wing being deflected downwards.
- NOT TRUE that the top of the wing acts as a constricting nozzle to speed up the airflow.

These theories are seen all over the internet but they do not paint a complete picture. The reason they live on is that they all carry some truth but not the whole truth. For example, in the first lift lie the air typically does flow faster over the top of the wing but it will not necessarily *meet up* with the air flowing under the wing. Regarding the second lift lie, only at extremely high altitude flights at high velocities (e.g. a spaceship re-entering earth's atmosphere at an altitude of 50 miles and speed of 10,000 mph) will the air deflected downwards by the bottom of the wing give a good estimate of lift. In the third lift lie, once again the air does speed up but there is no constriction as there is no other surface; when applied, this model predicts incorrect fluid speeds.

One *correct* way to describe lift is as follows:

Lift is a force that results from the turning of a moving fluid.

Coanda Effect vs Venturi Effect vs Bernoulli Principle
Disclaimer: some of the demos described here can be tricky to explain. In many instances, physicists erroneously explain them simply by saying "This is an example of the Bernoulli principle." In general, even these explanations are simplified. It suffices to say that if you go on and take advanced fluid dynamics courses you can learn more about the full explanation to these topics. That said, they are so cool and so commonly discussed in physics courses it is appropriate to spend a little time on them and have some fun!

The Coanda effect states that moving fluid will follow a curved surface causing the fluid flow to turn. Examples include blowing air past some hanging light bulbs, blowing air over a curved piece of paper, or blowing air over the top of a toilet paper roll on a paint roller, blowing air down a funnel at a ping pong ball, or the famous ball levitating in an airstream. Note: in some of these demos the physics, notably the last two, vortex shedding can occur complicating the fluid flow and explanation.

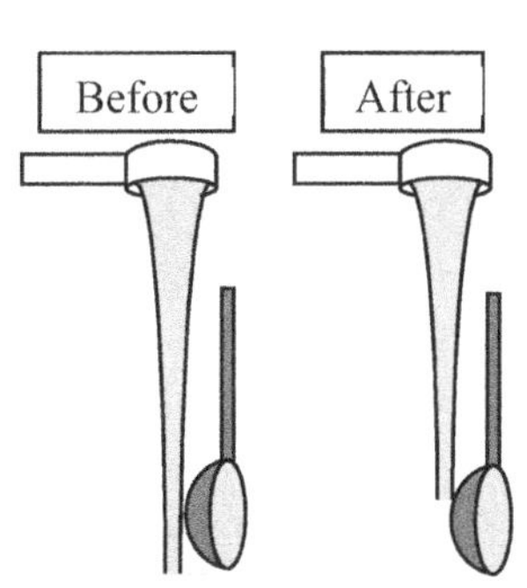

14.52 Consider the spoon shown at right slowly moved laterally so the convex portion of the spoon barely touches the fluid from the faucet. The water comes out of the faucet with speed v.

a) What is the pressure in the water stream? Does it depend on depth? Explain.
b) Consider the water flowing downwards from the faucet. Why does the stream become narrower as it falls? Hint: consider gravity/Bernoulli and the continuity equation.
c) Which way will the fluid be deflected <u>by the spoon</u>? Sketch it in the "After" figure.
d) Which direction is the force exerted on the spoon <u>by the fluid</u>?

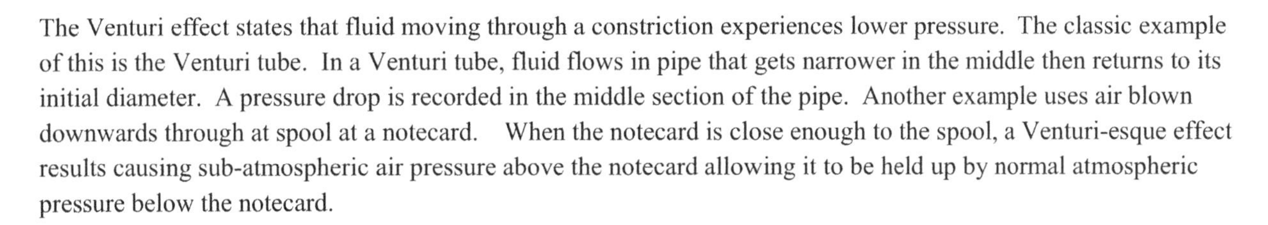

The Venturi effect states that fluid moving through a constriction experiences lower pressure. The classic example of this is the Venturi tube. In a Venturi tube, fluid flows in pipe that gets narrower in the middle then returns to its initial diameter. A pressure drop is recorded in the middle section of the pipe. Another example uses air blown downwards through at spool at a notecard. When the notecard is close enough to the spool, a Venturi-esque effect results causing sub-atmospheric air pressure above the notecard allowing it to be held up by normal atmospheric pressure below the notecard.

From the Bernoulli principle we know a non-viscous flow at constant depth experiencing an increase in fluid speed causes simultaneously exhibits a decrease in fluid pressure (in the absence of mechanical work being done on the fluid). Notice the Venturi effect is directly related to the Bernoulli principle but the Coanda effect is something different.

LEVITATING BALL

Many people (including physicists) have seen this demo explained in school. Most physics teachers at the high school and undergrad level have heard the same thing: "The blower increases fluid speed which causes the pressure to drop. The non-moving air outside the air stream has higher pressure and the ball is pushed from higher pressure to low. This is the force pushing the ball back into the airstream."

Sounds good, right?

Many people erroneously believe the pressure in the airstream produced by the blower is lower than atmospheric pressure because it is moving faster. While it is moving faster, it is not moving faster due to a constriction as in the Venturi effect. It moves faster because a blower has done mechanical work on it. If you build a pressure gauge and correctly test the pressure in the blower's air stream you will find the moving air is at atmospheric pressure!

That said, the air stream forced out of the blower is gradually diverging and is creating turbulence. These factors do tend to lower the pressure in the airstream at some point, I would *guess*, but I don't think that is the main issue. Drag creates a upwards force on the ball in the direction of the fluid flow. If half of the ball is completely out of the airstream, the Coanda effect (deflecting air flows) could probably be used to model a restoring force perpendicular to the fluid flow. These forces combine to bring a ball from halfway out of the stream deeper into the angled air stream. But perhaps there is more (entrainment, vortex shedding, etc)?

A more complicated phenomena called vortex shedding is also involved in this demonstration. The flow around the ball is not laminar! Read more about this complex topic by doing a web search for "vortex shedding levitating ball". There are some kick-ass simulations and the first several links are all interesting.

You might also read about entrainment. Imagine a stream of flowing colored water moving across the surface of a pool filled with clear water. As the colored water stream flows it will tend to drag along some of the non-colored water from the pool. This process is called entrainment. Entrainment will also affect a levitating ball system.

In closing, incorrect explanations of fluid demos are common. Just because an explanation is easy to understand does not mean it is correct. Try to find a reputable source such as a professional engineer or physicist specializing in fluid dynamics. Don't be afraid to admit you lack full understanding of these topics…I'm still pretty clueless about fluids after researching for many days and teaching for many years.

This is the perfect time to ask "Do you know how a parachute works?"

Web search for "bad lift demonstrations".

Turns out a parachute and the levitating ball have many physics concepts in common. Perhaps ask yourself why you can levitate a large beach ball in an airstream but not a fluffy cotton ball or crumpled up ball of paper? Have fun!

Glass of water trick
Put some cheese cloth (I use a double layer) or a fine wire mesh (like screen for a screen door) over a jar or glass. Secure the mesh in place with several rubber bands or the lid to the jar. Set the jar on the table in front of you and hopefully your students won't notice the mesh.

Pour water in the jar through the mesh. Place a card completely covering the lid on top of the jar. While holding the card in place, invert the jar. Remove your hand and the card stays in place.

Ask a student to come sit under the glass. Give them a poncho if your technique is terrible. Hold the glass over a student's head and slowly slide the card out so there is only the mesh holding up the water! A combination of surface tension and atmospheric pressure are at play once the card is removed.

Fun things to try on your own

- What is the largest mesh you can use?
- What is the largest jar/bowl you can use?
- Try it with no mesh and determine the largest jar/bowl you can use?
- With no mesh in place, is it easier, harder, or the same difficulty to do the demo when the glass is full, almost full, half full, almost empty, etc?

Page intentionally left blank.
This allows me to add content later if needed without having to re-format the cover.

Page intentionally left blank.

Page intentionally left blank.

Page intentionally left blank.

Page intentionally left blank.

Equation Sheet

$V_{sphere} = \frac{4}{3}\pi R^3$	$V_{box} = LWH$	$V_{cyl} = \pi R^2 H$	$\rho = \frac{M}{V}$
$A_{sphere} = 4\pi R^2$	$V = (A_{base}) \times (height)$	$A_{circle} = \pi R^2$	$x = \frac{-b \pm \sqrt{b^2 - 4ac}}{2a}$
$C = 2\pi R$	$A_{rect} = LW$	$A_{CylSide} = 2\pi RH$	
$160\underline{9}$ m = 1 mi	12 in = 1 ft	60 s = 1 min	1000 g = 1 kg
2.54 cm = 1 in	1 cc = 1 cm^3 = 1 mL	60 min = 1 hr	100 cm = 1 m
1 cm = 10 mm	1 yard = 3 ft	3600 s = 1 hr	1 km = 1000 m
1 furlong = 220 yards	$528\underline{0}$ ft = 1 mi	24 hrs = 1 day	1 rev = 2π rad = 360°
$g = 9.8 \frac{\text{m}}{\text{s}^2}$	$G = 6.67 \times 10^{-11} \frac{\text{N} \cdot \text{m}^2}{\text{kg}^2}$	$P_0 = 1.0 \times 10^5$ Pa	1 eV = $1.60\underline{2} \times 10^{-19}$ J
$1\text{ N} = 1 \frac{\text{kg} \cdot \text{m}}{\text{s}^2}$	1 J = 1 N · m	$1\text{ Pa} = 1 \frac{\text{N}}{\text{m}^2}$	
$x_f = x_i + v_{ix}t + \frac{1}{2}a_x t^2$	$v_{fx}^2 = v_{ix}^2 + 2a_x(\Delta x)$	$v_{fx} = v_{ix} + a_x t$	$r = \sqrt{x^2 + y^2}$
$\vec{A} \cdot \vec{B} = AB \cos\theta_{AB}$	$\lVert\vec{A} \times \vec{B}\rVert = AB \sin\theta_{AB}$	$\sin(A \pm B)$ $= \sin A \cos B \pm \cos A \sin B$	$\cos(A \pm B)$ $= \cos A \cos B \mp \sin A \sin B$
$\vec{v}_{ae} + \vec{v}_{eb} = \vec{v}_{ab}$	$\hat{r} = \cos\theta\,\hat{\imath} + \sin\theta\,\hat{\jmath}$	$\hat{\theta} = -\sin\theta\,\hat{\imath} + \cos\theta\,\hat{\jmath}$	
$a_{tan} = r\alpha$	$a_c = \frac{v^2}{r} = r\omega^2$	$\vec{a} = a_r\hat{r} + a_{tan}\hat{\theta}$	$\vec{a} = a_c(-\hat{r}) + a_{tan}\hat{\theta}$
$\Sigma\vec{F} = m\vec{a}$	$f \leq \mu n$	$F_G = \frac{GmM}{r^2}(-\hat{r})$	$U_G = -\frac{GmM}{r}$
$TKE = \frac{1}{2}mv^2$	$RKE = \frac{1}{2}I\omega^2$	$U_S = SPE = \frac{1}{2}kx^2$	$U_G = GPE = mgh$
$E_i + W_{non-con\ or\ ext} = E_f$	$\Delta KE = W_{ext.\&\,non-con}$	$W = Fd\cos\theta = F_{\parallel}d$	$W = \int F_x dx$
$\Delta U = -W = -\int_i^f \vec{F} \cdot d\vec{s}$	$F_x = -\frac{d}{dx}U(x)$	$\mathcal{P}_{inst} = \frac{dE}{dt} = \vec{F} \cdot \vec{v}$	$\mathcal{P}_{avg} = \frac{\Delta E}{\Delta t} = \frac{Work}{time}$
$\vec{J} = \Delta\vec{p} = \vec{F}\Delta t$	$\vec{p} = m\vec{v}$	$x_{CM} = \frac{m_1 x_1 + m_2 x_2}{m_1 + m_2}$	$x_{CM} = \frac{\int x\,dm}{\int dm}$
$\vec{\tau} = \vec{r} \times \vec{F}$	$\Sigma\vec{\tau} = I\vec{\alpha}$	$L = I\omega = mvr_{\perp}$	$\mathcal{P}_{inst} = \vec{\tau} \cdot \vec{\omega}$
$s = r\Delta\theta$	$v = r\omega$	$a_{tan} = r\alpha$	$a_c = \frac{v^2}{r} = r\omega^2$
$I_{\parallel axis} = I_{CM} + md^2$	$I_{zz} = I_{xx} + I_{yy}$	$I = \int r^2 dm$	$\frac{F}{A} = E\frac{\Delta L}{L_0}$
$P = \frac{F}{A}$	$P_{gauge} = P_{abs} - P_{ambient}$	$B = \rho_f V_{disp} g$	$A_1 v_1 = A_2 v_2$
$P(h) = P_0 + \rho gh$	$P + \frac{1}{2}\rho v^2 + \rho gh = \text{constant}$	$R = \frac{\pi r^4 \Delta P}{8\eta L}$	$F = \eta A \frac{\Delta v_x}{\Delta y}$

Prefix	**Abbreviation**	**$10^?$**		**Prefix**	**Abbreviation**	**$10^?$**
Giga	G	10^9		milli	m	10^{-3}
Mega	M	10^6		micro	μ	10^{-6}
kilo	k	10^3		nano	n	10^{-9}
centi	c	10^{-2}		pico	p	10^{-12}
				femto	f	10^{-15}

Page intentionally left blank.

Made in the USA
Las Vegas, NV
14 February 2023

67538008R00090